Crop Nutrition
Principles and Practices

Crop Nutrition
Principles and Practices

Barun Guha

RANDOM

RANDOM PUBLICATIONS
NEW DELHI (INDIA)

Crop Nutrition: Principles and Practices

ISBN 978-93-5111-365-2

Published in 2014 in India by

RANDOM PUBLICATIONS

4376-A/4B, Gali Murari Lal, Ansari Road
New Delhi-110 002
Phone : +91-11-43580356, +91-11-23289044
e-mail: randomexports@gmail.com, sales@randompublications.com, info@randompublications.com

Reprinted 2023

Type Setting by: Friends Media, Delhi-110089
Digitally Printed at: Replika Press Pvt. Ltd.

Preface

Nutrition is the provision, to cells and organisms, of the materials necessary (in the form of food) to support life. Crops require a balanced diet of essential nutrients throughout their growth cycle. Yara is the world's leading producer of such nutrients through its wide range of nitrogen fertilizers. Many plant foods can be found in the soil, but often in insufficient quantities to sustain high crop yields. Soil and climatic conditions can also limit a plant's uptake of nutrients at key growth stages. The macronutrients are consumed in larger quantities and are present in plant tissue in quantities from 0.2% to 4.0%. Micro nutrients are present in plant tissue in quantities measured in parts per million, ranging from 5 to 200 ppm, or less than 0.02% dry weight. Most soil conditions across the world can provide plants with adequate nutrition and do not require fertilizer for a complete life cycle.

However, man can artificially modify soil through the addition of fertilizer to promote vigorous growth and increase yield. The plants are able to obtain their required nutrients from the fertilizer added to the soil. A colloidal carbonaceous residue, known as humus, can serve as a nutrient reservoir. Besides lack of water and sunshine, nutrient deficiency is a major growth limiting factor. Although nitrogen is plentiful in the Earth's atmosphere, relatively few plants engage in nitrogen fixation (conversion of atmospheric nitrogen to a biologically useful form). Most plants therefore require nitrogen compounds to be present in the soil in which they grow. Plant nutrition is a difficult subject to understand completely, partially because of the variation between different plants and even between different species or individuals of a given clone. Elements present at low levels may cause deficiency symptoms, and toxicity is possible at levels that are too high. Furthermore, deficiency of one element may present as symptoms of toxicity from another element, and vice versa. Carbon and oxygen are absorbed from the air, while other nutrients are absorbed from the soil. Green plants obtain their carbohydrate supply from the carbon dioxide in the air by the process of photosynthesis. Each of these nutrients is used in a different place for a different essential function. The burgeoning demand on the world food supply, coupled with concern over the use of chemical fertilizers, has led to an accelerated interest in the practice of precision agriculture. This practice involves the careful control and monitoring of plant nutrition to maximize the rate of growth and yield of crops, as well as their nutritional value.

This book is of great value to growers, agricultural consultants, agronomists, and plant scientists, as it provides a practical easy-to-use reference for determining, monitoring, and improving the nutritional needs of plants. The graphical presentations of plant interactions with nutrients and beneficial elements, and the straightforward explanations of how nutrient deficiencies arise are especially useful to those seeking knowledge of plant nutrition.

I thank all members of my team who have helped in the preparation of the book. My special thanks go to "Random Publications" who have published the book.

—Barun Guha

Contents

1

Plant Nutrition

Plants use inorganic minerals for nutrition, whether grown in the field or in a container. Complex interactions involving weathering of rock minerals, decaying organic matter, animals, and microbes take place to form inorganic minerals in soil. Roots absorb mineral nutrients as ions in soil water. Many factors influence nutrient uptake for plants. Ions can be readily available to roots or could be "tied up" by other elements or the soil itself. Soil too high in pH (alkaline) or too low (acid) makes minerals unavailable to plants.

Fertility or Nutrition

The term "fertility" refers to the inherent capacity of a soil to supply nutrients to plants in adequate amounts and in suitable proportions. The term "nutrition" refers to the interrelated steps by which a living organism assimilates food and uses it for growth and replacement of tissue. Previously, plant growth was thought of in terms of soil fertility or how much fertilizer should be added to increase soil levels of mineral elements. Most fertilizers were formulated to account for deficiencies of mineral elements in the soil. The use of soilless mixes and increased research in nutrient cultures and hydroponics as well as advances in plant tissue analysis have led to a broader understanding of plant nutrition. Plant nutrition is a term that takes into account the interrelationships of mineral elements in the soil or soilless solution as well as their role in plant growth. This interrelationship involves a complex balance of mineral elements essential and beneficial for optimum plant growth.

Essential Versus Beneficial

The term essential mineral element (or mineral nutrient) was proposed by Arnon and Stout (1939). They concluded three criteria must be met for an element to be considered essential. These criteria are: 1. A plant must be unable to complete its life cycle in the absence of the mineral element. 2. The function of the element must not be replaceable by another mineral element. 3. The element must be directly involved in plant metabolism. These criteria are important guidelines for plant nutrition but exclude beneficial mineral elements. Beneficial elements are those that can compensate for toxic effects of other elements or may replace mineral nutrients in some other less specific function such as the maintenance of osmotic pressure. The omission of beneficial nutrients in commercial production could mean that plants are not being grown to their optimum genetic potential but are merely produced at a subsistence level. This discussion of plant nutrition includes both the essential and beneficial mineral elements.

What are the Mineral Elements?

There are actually 20 mineral elements necessary or beneficial for plant growth. Carbon (C), hydrogen (H), and oxygen (O) are supplied by air and water. The six macronutrients, nitrogen (N), phosphorus (P), potassium (K), calcium (Ca), magnesium (Mg), and sulphur (S) are required by plants in large amounts. The rest of the elements are required in trace amounts (micronutrients). Essential trace elements include boron (B), chlorine (Cl), copper (Cu), iron (Fe), manganese (Mn), sodium (Na), zinc (Zn), molybdenum (Mo), and nickel (Ni). Beneficial mineral elements include silicon (Si) and cobalt (Co). The beneficial elements have not been deemed essential for all plants but may be essential for some. The distinction between beneficial and essential is often difficult in the case of some trace elements. Cobalt for instance is essential for nitrogen fixation in legumes. It may also inhibit ethylene formation (Samimy, 1978) and extend the life of cut roses (Venkatarayappa et al., 1980). Silicon, deposited in cell walls, has been found to improve heat and drought tolerance and increase resistance to insects and fungal infections. Silicon, acting as a beneficial element, can help compensate for toxic levels of manganese, iron, phosphorus and aluminium as well as zinc deficiency. A more holistic approach to plant nutrition would not be limited to nutrients essential to survival but would include mineral elements at levels beneficial for optimum growth. With developments in analytical chemistry and the ability to eliminate contaminants in nutrient cultures, the list of essential elements may well increase in the future.

The Mineral Elements in Plant Production

The use of soil for greenhouse production before the 1960s was common. Today a few growers still use soil in their mixes. The bulk of production is in soilless mixes. Soilless mixes must provide support, aeration, nutrient and moisture retention just as soils do, but the addition of fertilizers or nutrients are different. Many soilless mixes have calcium, magnesium, phosphorus, sulphur, nitrogen, potassium and some micronutrients incorporated as a pre-plant fertilizer. Nitrogen and potassium still must be applied to the crop during production. Difficulty in blending a homogenous mix using pre-plant fertilizers may often result in uneven crops and possible toxic or deficient levels of nutrients. Soilless mixes that require addition of micro and macronutrients applied as liquid throughout the growth of the crop, may actually give the grower more control of his crop. To achieve optimum production, the grower can adjust nutrient levels to compensate for other environmental factors during the growing season. The absorption of mineral ions is dependent on a number of factors in addition to weather conditions. These include the cation exchange capacity or CEC and the pH or relative amount of hydrogen (H+) or hydroxyl ions (OH-) of the growing medium, and the total alkalinity of the irrigation water.

Cec or Cation Exchange Capacity

The Cation Exchange Capacity refers to the ability of the growing medium to hold exchangeable mineral elements within its structure. These cations include ammonium nitrogen, potassium, calcium, magnesium, iron, manganese, zinc and copper. Peat moss and mixes containing bark, sawdust and other organic materials all have some level of cation exchange capacity.

Ph: What Does It Mean?

The term pH refers to the alkalinity or acidity of a growing media water solution. This solution consists of mineral elements dissolved in ionic form in water. The reaction of this solution whether it is acid, neutral or alkaline will have a marked effect on the availability of mineral elements to plant roots. When there is a greater amount of hydrogen H+ ions the solution will be acid (<7.0). If there is more hydroxyl OH- ions the solution will be alkaline (>7.0). A balance of hydrogen to hydroxyl ions yields a pH neutral soil (=7.0). The range for most crops is 5.5 to 6.2 or slightly acidic. This creates the greatest average level for availability for all essential plant nutrients. Extreme fluctuations of higher or lower pH can cause deficiency or toxicity of nutrients.

The Elements of Complete Plant Nutrition

The following is a brief guideline of the role of essential and beneficial mineral nutrients that are crucial for growth. Eliminate any one of these elements, and plants will display abnormalities of growth, deficiency symptoms, or may not reproduce normally.

Macronutrients

Nitrogen is a major component of proteins, hormones, chlorophyll, vitamins and enzymes essential for plant life. Nitrogen metabolism is a major factor in stem and leaf growth (vegetative growth). Too much can delay flowering and fruiting. Deficiencies can reduce yields, cause yellowing of the leaves and stunt growth.

Phosphorus is necessary for seed germination, photosynthesis, protein formation and almost all aspects of growth and metabolism in plants. It is essential for flower and fruit formation. Low pH (<4) results in phosphate being chemically locked up in organic soils. Deficiency symptoms are purple stems and leaves; maturity and growth are retarded. Yields of fruit and flowers are poor. Premature drop of fruits and flowers may often occur. Phosphorus must be applied close to the plant's roots in order for the plant to utilize it. Large applications of phosphorus without adequate levels of zinc can cause a zinc deficiency.

Potassium is necessary for formation of sugars, starches, carbohydrates, protein synthesis and cell division in roots and other parts of the plant. It helps to adjust water balance, improves stem rigidity and cold hardiness, enhances flavour and colour on fruit and vegetable crops, increases the oil content of fruits and is important for leafy crops. Deficiencies result in low yields, mottled, spotted or curled leaves, scorched or burned look to leaves.

Sulphur is a structural component of amino acids, proteins, vitamins and enzymes and is essential to produce chlorophyll. It imparts flavour to many vegetables. Deficiencies show as light green leaves. Sulphur is readily lost by leaching from soils and should be applied with a nutrient formula. Some water supplies may contain Sulphur.

Magnesium is a critical structural component of the chlorophyll molecule and is necessary for functioning of plant enzymes to produce carbohydrates, sugars and fats. It is used for fruit and nut formation and essential for germination of seeds. Deficient plants appear chlorotic, show yellowing between veins of older leaves; leaves may droop. Magnesium is leached by watering and must be supplied when feeding. It can be applied as a foliar spray to correct deficiencies.

Calcium activates enzymes, is a structural component of cell walls, influences water movement in cells and is necessary for cell growth and division. Some plants must have calcium to take up nitrogen and other minerals. Calcium is easily leached. Calcium, once deposited in plant tissue, is immobile (non-translocatable) so there must be a constant supply for growth. Deficiency causes stunting of new growth in stems, flowers and roots. Symptoms range from distorted new growth to black spots on leaves and fruit. Yellow leaf margins may also appear.

Micronutrients

Iron is necessary for many enzyme functions and as a catalyst for the synthesis of chlorophyll. It is essential for the young growing parts of plants. Deficiencies are pale leaf colour of young leaves followed by yellowing of leaves and large veins. Iron is lost by leaching and is held in the lower portions of the soil structure. Under conditions of high pH (alkaline) iron is rendered unavailable to plants. When soils are alkaline, iron may be abundant but unavailable. Applications of an acid nutrient formula containing iron chelates, held in soluble form, should correct the problem.

Manganese is involved in enzyme activity for photosynthesis, respiration, and nitrogen metabolism. Deficiency in young leaves may show a network of green veins on a light green background similar to an iron deficiency. In the advanced stages the light green parts become white, and leaves are shed. Brownish, black, or grayish spots may appear next to the veins. In neutral or alkaline soils plants often show deficiency symptoms. In highly acid soils, manganese may be available to the extent that it results in toxicity.

Boron is necessary for cell wall formation, membrane integrity, calcium uptake and may aid in the translocation of sugars. Boron affects at least 16 functions in plants. These functions include flowering, pollen germination, fruiting, cell division, water relationships and the movement of hormones. Boron must be available throughout the life of the plant. It is not translocated and is easily leached from soils. Deficiencies kill terminal buds leaving a rosette effect on the plant. Leaves are thick, curled and brittle. Fruits, tubers and roots are discoloured, cracked and flecked with brown spots.

Zinc is a component of enzymes or a functional cofactor of a large number of enzymes including auxins (plant growth hormones). It is essential to carbohydrate metabolism, protein synthesis and internodal elongation (stem growth). Deficient plants have mottled leaves with irregular chlorotic areas. Zinc deficiency leads to iron deficiency causing similar symptoms. Deficiency occurs on eroded soils and is least available at a pH range of 5.5 - 7.0. Lowering the pH can render zinc more available to the point of toxicity.

Copper is concentrated in roots of plants and plays a part in nitrogen metabolism. It is a component of several enzymes and may be part of the enzyme systems that use carbohydrates and proteins. Deficiencies cause die back of the shoot tips, and terminal leaves develop brown spots. Copper is bound tightly in organic matter and may be deficient in highly organic soils. It is not readily lost from soil but may often be unavailable. Too much copper can cause toxicity.

Molybdenum is a structural component of the enzyme that reduces nitrates to ammonia. Without it, the synthesis of proteins is blocked and plant growth ceases. Root nodule (nitrogen fixing) bacteria also require it. Seeds may not form completely, and nitrogen deficiency may occur if plants are lacking molybdenum. Deficiency signs are pale green leaves with rolled or cupped margins.

Chlorine is involved in osmosis (movement of water or solutes in cells), the ionic balance necessary for plants to take up mineral elements and in photosynthesis. Deficiency symptoms include wilting, stubby roots, chlorosis (yellowing) and bronzing. Odors in some plants may be decreased. Chloride, the ionic form of chlorine used by plants, is usually found in soluble forms and is lost by leaching. Some plants may show signs of toxicity if levels are too high.

Nickel has just recently won the status as an essential trace element for plants according to the Agricultural Research Service Plant, Soil and Nutrition Laboratory in Ithaca, NY. It is required for the enzyme urease to break down urea to liberate the nitrogen into a usable form for plants. Nickel is required for iron absorption. Seeds need nickel in order to germinate. Plants grown without additional nickel will gradually reach a deficient level at about the time they mature and begin reproductive growth. If nickel is deficient plants may fail to produce viable seeds.

Sodium is involved in osmotic (water movement) and ionic balance in plants. Cobalt is required for nitrogen fixation in legumes and in root nodules of nonlegumes. The demand for cobalt is much higher for nitrogen fixation than for ammonium nutrition. Deficient levels could result in nitrogen deficiency symptoms.

Silicon is found as a component of cell walls. Plants with supplies of soluble silicon produce stronger, tougher cell walls making them a mechanical barrier to piercing and sucking insects. This significantly enhances plant heat and drought tolerance. Foliar sprays of silicon have also shown benefits reducing populations of aphids on field crops. Tests have also found that silicon can be deposited by the plants at the site of infection by fungus to combat the penetration of the cell walls by the attacking fungus. Improved leaf erectness, stem strength and prevention or depression of iron and manganese toxicity have all been noted as effects from silicon. Silicon has not been determined essential for all plants but may be beneficial for many.

Physiographic factors of watershed consist of both, the watershed as with as channel characteristics.

Imbalances

Boron Deficiency (Plant Disorder)

Boron (B) deficiency is the most widespread micronutrient deficiency around the world and causes large losses in crop production both quantitatively and qualitatively. Boron deficiency affects vegetative and reproductive growth of plants resulting in inhibition of cell expansion, death of meristem and reduced fertility.

Plants contain Boron both in a water-soluble and insoluble form. In intact plants, the amount of water-soluble boron fluctuates with the amount of boron supplied, while insoluble boron does not. The appearance of a boron deficiency coincides with the decrease of water-insoluble boron. It appears that the insoluble boron is the functional form while the soluble boron represents the surplus.

Boron is essential for the growth of higher plants. The primary function of the element is to provide structural integrity to the cell wall in plants. Other functions likely include the maintenance of the plasma membrane and other metabolic pathways.

Symptoms

Symptoms include dying growing tips and bushy stunted growth, extreme cases may prevent fruit set. Crop-specific symptoms include;

- *Apple*- interacting with calcium, may display as "water core", internal areas appearing frozen
- *Beetroot*- rough, cankered patches on roots, internal brown rot.
- *Cabbage*- distorted leaves, hollow areas in stems.
- *Cauliflower*- poor development of curds, and brown patches. Stems, leafstalks and midribs roughened.
- *Celery*- leaf stalks develop cracks on the upper surface, inner tissue is reddish brown.

- *Celeriac*- causes brown heart rot
- *Pears*- new shoots die back in spring, fruits develop hard brown flecks in the skin.
- *Strawberries*- Stunted growth, foliage small, yellow and puckered at tips. Fruits are small and pale.
- *Swede (rutabaga)* and *turnip*- brown or gray concentric rings develop inside the roots.
- *Arecaceae* (*Palm Tree*) - brown spots on fronds & lower productivity.references/

Soil Conditions

Boron is present in the soil in many form, the most common being Boric Acid (H_3BO_3). An adequate amount of boron in the soil is 12 mg/kg. If the boron content of the soil drops below 0.14 mg/kg then boron deficiency is likely to be observed. Boron deficiency is also observed in basic soils with a high pH because in basic conditions boric acid exists in an undissociated form which the plant is unable to absorb. Soils with low organic matter content (<1.5%) are also susceptible to boron deficiency. Highly leached sandy soils are also characteristic of boron deficiency because the boron will not be retained in the soil. Boron toxicity is also possible if the boron content of the soil is high enough that the plant cannot cope with the excess boron. The levels at which boron is toxic to plants varies with different species of plants.

Boron Requirements

Boron is an essential micronutrient which means it is essential for plant growth and development, but is required in very small quantities. Although Boron requirements vary among crops, the optimum boron content of the leaves for most crops is 20-100 ppm. Excess boron can result in boron toxicity and the toxicity level varies between plants.

Treatment

Boric acid (16.5%boron), borax (11.3% boron) or SoluBor (20.5% boron) can be applied to soils to correct boron deficiency. Typical applications of actual boron are about 1.1 kg/hectare or 1.0 lb/acre but optimum levels of boron vary with plant type. Borax, Boric Acid or Solubor can be dissolved in water and sprayed or applied to soil as a dust. Excess boron is toxic to plants so care must be taken to ensure correct application rate and even coverage. Leaves of many plants are damaged by boron; therefore, when in doubt, only apply to soil.

Application of boron may not correct boron deficiency in alkaline soils because even with the addition of boron, it may remain unavailable for plant absorption. Continued application of boron may be necessary in soils that are susceptible to leaching such as sandy soils. Flushing soils containing toxic levels of boron with water can remove the boron through leaching.

Functions

Once boron has been absorbed by the plant and incorporated into the various structures that require boron, it is unable to disassemble these structures and re-transport boron through the plant resulting in boron being a non-mobile nutrient. Due to translocation difficulties the youngest leaves often show deficiency symptoms first.

Cell Wall

Boron is part of the dRG-II-B complex which is involved in the cross linking for pectin located in the primary cell wall and the middle lamella of plant cells. This cross linking is thought to stabilize the matrix of plant cell walls.

Calcium Deficiency (Plant Disorder)

Calcium (Ca) deficiency is a plant disorder that can be caused by insufficient calcium in the growing medium, but is more frequently a product of low transpiration of the whole plant or more commonly the affected tissue. Plants are susceptible to such localized calcium deficiencies in low or nontranspiring tissues because calcium is not transported in the phloem. This may be due to water shortages, which slow the transportation of calcium to the plant, poor uptake of calcium through the stem, or can be caused by excessive usage of potassium or nitrogen fertilizers.

Causes

Acidic, sandy, or coarse soils often contain less calcium. Uneven soil moisture and overuse of fertilizers can also cause calcium deficiency. At times, even with sufficient calcium in the soil, it can be in an insoluble form and is then unusable by the plant or it could be attributed to a "transport protein". Soils containing high phosphorus are particularly susceptible to creating insoluble forms of calcium.

Symptoms

Calcium deficiency symptoms appear initially as localized tissue necrosis leading to stunted plant growth, necrotic leaf margins on young leaves or curling of the leaves, and eventual death of terminal buds and root tips. Generally, the new growth and rapidly growing tissues of the plant are affected first. The mature leaves are rarely if ever affected because calcium accumulates to high concentrations in older leaves.

Figure: *Blossom end rot on a grape tomato*

Crop-specific Symptoms Include:

Apple: 'Bitter pit' – fruit skins develop pits, brown spots appear on skin and/or in flesh and taste of those areas is bitter. This usually occurs when fruit is in storage, and Bramley apples are particularly susceptible. Related to boron deficiency, "water cored" apples seldom display bitter pit effects.

Cabbage and Brussels Sprouts: Internal browning and "tip burn"

Carrot: 'Cavity spot' – oval spots develop into craters which may be invaded by other disease-causing organisms.

Celery: Stunted growth, central leaves stunted.

Tomatoes and Peppers: 'Blossom end rot' – Symptoms start as sunken, dry decaying areas at the blossom end of the fruit, furthest away from the stem, not all fruit on a truss is necessarily affected. Sometimes rapid growth from high-nitrogen fertilizers may exacerbate blossom end rot.

Treatment

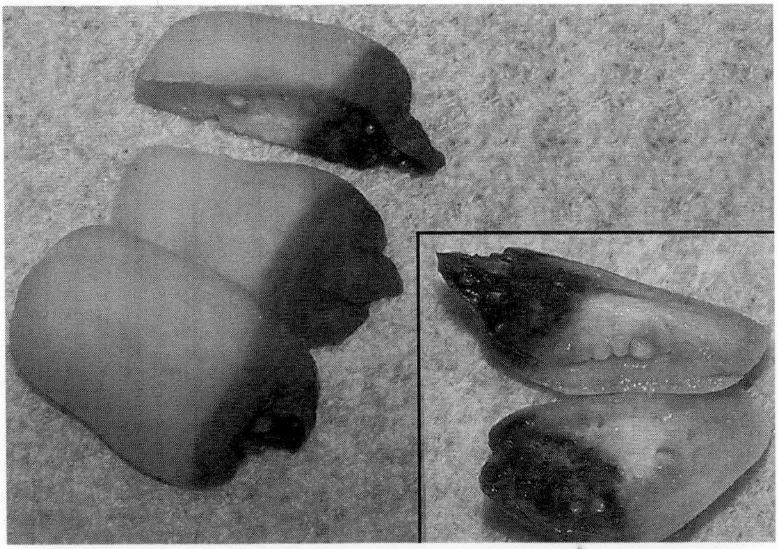

Figure: *Dissection of grape tomato with blossom end rot*

Calcium deficiency can sometimes be rectified by adding agricultural lime to acid soils, aiming at a pH of 6.5, unless the subject plants specifically prefer acidic soil. Organic matter should be added to the soil to improve its moisture-retaining capacity. However, because of the nature of the disorder (i.e. poor transport of calcium to low transpiring tissues), the problem cannot generally be cured by the addition of calcium to the roots. In some species, the problem can be reduced by prophylactic spraying with calcium chloride of tissues at risk.

Plant damage is difficult to reverse, so corrective action should be taken immediately, supplemental applications of calcium nitrate at 200 ppm nitrogen, for example. Soil pH should be tested, and corrected if needed, because calcium deficiency is often associated with low pH.

Iron Deficiency (Plant Disorder)

Iron (Fe) deficiency is a plant disorder also known as "lime-induced chlorosis". It can be confused with manganese deficiency. A deficiency in the soil is rare but iron can be unavailable for absorption if soil pH is not between about 5 and 6.5. A common problem is when the soil is too alkaline (the pH is above 6.5). Also, iron deficiency can develop if the soil is too waterlogged or has been overfertilized. Elements like calcium, zinc, manganese, phosphorus, or copper can tie up iron if they are present in high amounts.

Iron is needed to produce chlorophyll, hence its deficiency causes chlorosis. For example, iron is used in the active site of glutamyl-tRNA reductase, an enzyme needed for the formation of 5-Aminolevulinic acid which is a precursor of heme and chlorophyll.

Symptoms

Symptoms include leaves turning yellow or brown in the margins between the veins which may remain green, while young leaves may appear to be bleached. Fruit would be of poor quality and quantity. Any plant may be affected, but raspberries and pears are particularly susceptible, as well as most acid-loving plants such as azaleas and camellias.

Treatment

Iron deficiency can be avoided by choosing appropriate soil for the growing conditions (e.g., avoid growing acid loving plants on lime soils), or by adding well-rotted manure or compost. If iron deficit chlorosis is suspected then check the pH of the soil with an appropriate test kit or instrument. Take a soil sample at surface and at depth. If the pH is over seven then consider soil remediation that will lower the pH toward the 6.5 - 7 range. Remediation includes: i) adding compost, manure, peat or similar organic matter (warning. Some retail blends of manure and compost have pH in the range 7 - 8 because of added lime. Read the MSDS if available. Beware of herbicide residues in manure. Source manure from a certified organic source.) ii) applying Ammonium Sulphate as a Nitrogen fertilizer (acidifying fertilizer due to decomposition of ammonium ion to nitrate in the soil and root zone) iii) applying elemental Sulphur to the soil (oxidizes over the course of months to produce sulphate/sulphite and lower pH). Note: adding acid directly e.g. sulphuric/hydrochloric/citric acid is dangerous as you may mobilize metal ions in the soil that are toxic and otherwise bound. Iron can be made available immediately to the plant by the use of iron sulphate or iron chelate compounds. Two common iron chelates are Fe EDTA and Fe EDDHA. Iron sulphate (Iron(II)_sulphate) and iron EDTA are only useful in soil up to PH 7.1 but they can be used as a foliar spray (Foliar_feeding). Iron

EDDHA is useful up to PH 9 (highly alkaline) but must be applied to the soil and in the evening to avoid photodegradation. EDTA in the soil may mobilize Lead, EDDHA does not appear to.

Magnesium Deficiency (Plants)

Magnesium (Mg) deficiency is a detrimental plant disorder that occurs most often in strongly acidic, light, sandy soils, where magnesium can be easily leached away. Magnesium is an essential macronutrient found from 0.2-0.4% dry matter and is necessary for normal plant growth.

Figure: *A plant with Magnesium deficiency*

Excess potassium, generally due to fertilizers, further aggravates the stress from the magnesium deficiency. Magnesium has an important role in photosynthesis because it forms the central atom of chlorophyll. Therefore, without sufficient amounts of magnesium, plants begin to degrade the chlorophyll in the old leaves. This causes the main symptom of magnesium deficiency, chlorosis, or yellowing between leaf veins, which stay green, giving the leaves a marbled appearance. Due to magnesium's mobile nature, the plant will first break down chlorophyll in older leaves and transport the Mg to younger leaves which have greater photosynthetic needs. Therefore, the first sign of magnesium deficiency is the chlorosis of old leaves which progresses to the young leaves as the deficiency continues. Magnesium also is a necessary activator for many critical enzymes, including ribulosbiphosphate carboxylase (RuBisCO) and phosphoenolpyruvate carboxylase (PEPC), both essential enzymes in carbon fixation. Thus low amounts of Mg lead to a decrease in photosynthetic and enzymatic activity within the plants. Magnesium is also crucial in stabilizing ribosome structures, hence, a lack of magnesium causes depolymerization of ribosomes leading to pre-mature aging of the plant. After prolonged magnesium deficiency, necrosis and dropping of older leaves occurs. Plants deficient in magnesium also produce smaller, woodier fruits.

Magnesium deficiency may be confused with zinc or chlorine deficiencies, viruses, or natural aging since all have similar symptoms. Adding Epsom salts (diluted to 8.5 oz. per 2.2 gal. of water) or crushed dolomitic limestone to the soil can rectify magnesium deficiencies. For a more organic solution, applying home-made compost mulch can prevent leaching during excessive rainfall and provide plants with sufficient amounts of nutrients, including magnesium.

Manganese Deficiency (Plant)

Manganese (Mn) deficiency is a plant disorder that is often confused with, and occurs with, iron deficiency. Most common in poorly drained soils, also where organic matter levels are high. Manganese may be unavailable to plants where pH is high.

Affected plants include onion, apple, peas, French beans, cherry and raspberry, and symptoms include yellowing of leaves with smallest leaf veins remaining green to produce a 'chequered' effect. The plant may seem to grow away from the problem so that younger leaves may appear to be unaffected. Brown spots may appear on leaf surfaces, and severely affected leaves turn brown and wither.

Prevention can be achieved by improving soil structure. Do not over-lime.

Nitrogen Deficiency

All plants require sufficient supplies of macronutrients for healthy growth, and nitrogen (N) is a nutrient that is commonly in limited supply. Nitrogen deficiency in plants can occur when organic matter with high carbon content, such as sawdust, is added to soil. Soil organisms use any nitrogen to break down carbon sources, making N unavailable to plants. This is known as "robbing" the soil of nitrogen. All vegetables apart from nitrogen fixing legumes are prone to this disorder.

Nitrogen deficiency can be prevented in the short term by using grass mowings as a mulch, or foliar feeding with manure, and in the longer term by building up levels of organic matter in the soil. Sowing green manure crops such as grazing rye to cover soil over the winter will help to prevent nitrogen leaching, while leguminous green manures such as winter tares will fix additional nitrogen from the atmosphere.

Detection

The visual symptoms of nitrogen deficiency mean that it can be relatively easy to detect in some plant species. Symptoms include poor plant growth, and leaves that are pale green or yellow because they are unable to make sufficient chlorophyll. Leaves in this state are said to be chlorotic. Lower leaves (older leaves) show symptoms first, since the plant will move nitrogen from older tissues to more important younger ones. Nevertheless, plants are reported to show nitrogen deficiency symptoms at different parts. For example Nitrogen deficiency of tea is identified by retarded shoot growth and yellowing of younger leaves.

However, these physical symptoms can also be caused by numerous other stresses, such as deficiencies in other nutrients, toxicity, herbicide injury, disease, insect damage or environmental conditions. Therefore, nitrogen deficiency is most reliably detected by conducting quantitative tests in addition to assessing the plants visual symptoms. These tests include soil tests and plant tissue test.

Plant tissue tests destructively sample the plant of interest. However, nitrogen deficiency can also be detected non-destructively by measuring chlorophyll content.

Chlorophyll content tests work because leaf nitrogen content and chlorophyll concentration are closely linked, which would be expected since the majority of leaf nitrogen is contained in chlorophyll molecules. Chlorophyll content can be detected with a Chlorophyll content metre; a portable instrument that measures the greenness of leaves to estimate their relative chlorophyll concentration.

Chlorophyll content can also be assessed with a chlorophyll fluorometer, which measures a chlorophyll fluorescence ratio to identify phenolic compounds that are produced in higher quantities when nitrogen is limited. These instruments can therefore be used to non-destructively test for nitrogen deficiency.

Phosphorus Deficiency

Phosphorus deficiency is a plant disorder that is most common in areas of high rainfall, especially on acid, clay or poor chalk soils.

Cold weather can cause a temporary deficiency. All plants may be affected, although this is an uncommon disorder. Particularly susceptible are carrots, lettuce, spinach, apples, currants and gooseberries. Symptoms include poor growth, and leaves that turn blue/green but not yellow—oldest leaves are affected first. Fruits are small and acid tasting. Phosphorus deficiency may be confused with nitrogen deficiency. Undersides of tomato plant leaves, and the veins and stems, may turn purple. stiff, stunted plants with purlish tinge.

It can be controlled by applying organic sources of phosphorus such as rock phosphate or a high-phosphorus fish fertilizer applied as a soil drench or foliar feed.

Plants that are naturally adapted to low levels of available soil phosphorus, however, are more likely to suffer from phosphate poisoning: the key is to provide the right level for any particular plant type, neither too high nor too low.

Potassium Deficiency (Plants)

Soil Type and Potassium Deficiency

Potassium deficiency, also known as potash deficiency, is a plant disorder that is most common on light, sandy soils, because potassium ions (K^+) are highly soluble and will easily leach from soils without colloids. Potassium deficiency is also common in chalky or peaty soils with a low clay content. It is also found on heavy clays with a poor structure.

Role of Potassium in Plants

The main role of potassium is to provide the appropriate ionic environment for metabolic processes in the cytosol, and as such functions as a regulator of various processes including growth regulation. Plants require potassium ions (K^+) for protein synthesis and for the opening and closing of stomata, which is regulated by proton pumps to make surrounding guard cells either turgid or flaccid. A deficiency of potassium ions can impair a plant's ability to maintain these processes. Potassium also functions in other physiological processes such as photosynthesis, protein synthesis, activation of some enzymes, phloem solute transport of photoassimilates into source organs, and maintenance of cation: anion balance in the cytosol and vacuole.

Symptoms of Potassium Deficiency

Typical symptoms of potassium deficiency in plants include brown scorching and curling of leaf tips as well as chlorosis (yellowing) between leaf veins. Purple spots may also appear on the leaf undersides.

Plant growth, root development, and seed and fruit development are usually reduced in potassium-deficient plants. Often, potassium deficiency symptoms first appear on older (lower) leaves because potassium is a mobile nutrient, meaning that a plant can allocate potassium to younger leaves when it is K deficient. Deficient plants may be more prone to frost damage and disease, and their symptoms can often be confused with wind scorch or drought. The deficiency is most common in several important fruit and vegetable crops; notably potatoes, brassicas, tomatoes, apples, currants, gooseberries, and raspberries. Sugar beets, cereals, and clover are also commonly affected. Specific symptoms for each of these plants are as follows:

In potatoes, tuber size is much reduced and crop yield is low. The leaves of the plant appear dull and are often blue-green in colour with intervenal chlorosis. Leaves will also develop small, dark brown spots on the undersides and a bronzed appearance on the upper surfaces.

In brassicas, leaves are blue-green in colour and may have a low degree of intervenal chlorosis. Scorching along the outside edges of leaves is common, and leaves are often tough in texture due to slow growth.

In tomatoes, the stems are woody and growth is slow. Leaves are blue-green in colour, and the intervenal area often fades to a pale gray colour. Leaves may also have a bronzed appearance and yellow and orange patches may develop on some of the leaflets. Fruits often ripen unevenly and sometimes have green patches near the stalks.

In apples, leaves are scorched around the edges, and intervenal chlorosis is common. Apple fruits often have a slightly acidic or woody taste.

In gooseberries, currants, and raspberries, dieback of shoots and branches is common and although the plant may produce many blossom buds in the early stages of deficiency, fruit yields turn out low and the fruits are of poor quality.

Potassium Deficiency and Plant Disease

For many species, potassium-deficient plants are more susceptible to frost damage and certain diseases than plants with adequate potassium levels. Increased disease resistance associated with adequate potassium levels indicates that potassium has roles in providing disease resistance, and increasing the potassium levels of deficient plants have been shown to decrease the intensity of many diseases. However, increasing potassium concentration above the optimal level does not provide greater disease resistance. In agriculture, some cultivars are more efficient at K uptake due to genetic variations, and often these plants have increased disease resistance. The mechanisms involved with increased host resistance and potassium include a decreased cell permeability and decreased susceptibility to tissue penetration. Silica, which is accumulated in greater quantities when adequate potassium is present, is incorporated into cell walls, strengthening the epidermal layer which functions as a physical barrier to pathogens. Potassium has also been implicated to have a role in the proper thickening of cell walls.

Prevention and Cure

Prevention and cure can be achieved in the shorter term by adding potassium-specific fertilizer, often called potash, which consists of K_2CO_3. Rock potash may be a good solution because it has a high potassium content but is released slowly to reduce overdose. Common forms of inorganic fertilizers include potassium nitrate, potassium sulphate, and monopotassium phosphate. Other potassium-rich treatments include feeding with home-made comfrey liquid, adding seaweed meal, composted bracken, compost rich in decayed banana peels, or other organic potassium-rich fertilizers. Wood ash has high potassium content, but should be composted first as it is in a highly soluble form. Adequate moisture is necessary for effective potassium uptake; low soil water reduces K uptake by plant roots. Liming acidic soils will also increase K retention in the soil by reducing leaching. In the longer term the soil structure should be improved by adding plenty of well rotted compost or manure. Additionally, improving soil structure will help reduce leaching.

Micronutrient Deficiency

A micronutrient deficiency (or trace mineral deficiency) is a physiological plant disorder which occurs when a micronutrient is deficient in the soil in which a plant grows. Micronutrients are distinguished from macronutrients (such as nitrogen, phosphorus, and potassium) by the relatively low quantities needed by the plant. It also represents a human physiological disorder which occurs when a person's diet does not contain required nutrients and/or when illnesses (such as diarrhoea or malaria) cause rapid loss of nutrients through feces or vomit. Untreated, sub-clinical deficiencies can manifest as serious physical disabilities or life-threatening disorders, such as Beriberi (thiamine deficiency), scurvy (vitamin C) or xerophthalmia (vitamin A).

A number of elements are known to be needed in these small amounts for proper plant growth and development. Nutrient deficiencies in these areas can adversely affect plant growth and development. Some of the best known trace mineral deficiencies include: boron deficiency, calcium deficiency, iron deficiency, magnesium deficiency, and manganese deficiency.

Micronutrient Deficiency can also be found in people. Micronutrient deficiencies in children are associated with 10% of all children's deaths. Micronutrient Deficiencies are when a child is lacking essential vitamins or minerals such as Vitamin A, iron, and zinc. These deficiencies are caused by a long-term lack of nutritious food or they can be caused by infections such as worms.

List of Essential Trace Minerals for Plants

- Boron is believed to be involved in carbohydrate transport in plants; it also assists in metabolic regulation. Boron deficiency will often result in bud dieback.
- Calcium
- Chlorine is necessary for osmosis and ionic balance; it also plays a role in photosynthesis.
- Cobalt is essential to plant health. Cobalt is thought to be an important catalyst in nitrogen fixation. It may need to be added to some soils before seeding legumes.
- Copper is a component of some enzymes and of vitamin A. Symptoms of copper deficiency include browning of leaf tips and chlorosis.
- Iron is essential for chlorophyll synthesis, which is why an iron deficiency results in chlorosis.
- Manganese activates some important enzymes involved in chlorophyll formation. Manganese deficient plants will develop chlorosis between the veins of its leaves. The availability of manganese is partially dependent on soil pH.
- Molybdenum is essential to plant health. Molybdenum is used by plants to reduce nitrates into usable forms. Some plants use it for nitrogen fixation, thus it may need to be added to some soils before seeding legumes.
- Nickel
- Zinc participates in chlorophyll formation, and also activates many enzymes. Symptoms of zinc deficiency include chlorosis and stunted growth.

Chlorosis

In botany, chlorosis is a condition in which leaves produce insufficient chlorophyll. As chlorophyll is responsible for the green colour of leaves, chlorotic leaves are pale, yellow, or yellow-white. The affected plant has little or no ability to manufacture carbohydrates through photosynthesis and may die unless the cause of its chlorophyll insufficiency is treated, although some chlorotic plants, such as the albino *Arabidopsis thaliana* mutant *ppi2*, are viable if supplied with exogenous sucrose.

Chlorosis is derived from the word Chloris or from the Greek *khloros* meaning 'greenish-yellow', 'pale green', 'pale', 'pallid', or 'fresh'.

Causes

Figure: *A* Liquidambar *leaf with interveinal chlorosis*

Chlorosis is typically caused when leaves do not have enough nutrients to synthesise all the chlorophyll they need. It can be brought about by a combination of factors including:

- a specific mineral deficiency in the soil, such as iron or magnesium

- deficient nitrogen and/or proteins
- a soil pH at which minerals become unavailable for absorption by the roots
- poor drainage (waterlogged roots)
- damaged and/or compacted roots
- pesticides and particularly herbicides may cause chlorosis, both to target weeds and occasionally to the crop being treated.
- exposure to sulphur dioxide
- ozone injury to sensitive plants

However, the exact conditions vary from plant type to plant type. For example, Azaleas grow best in acidic soil and rice is unharmed by waterlogged soil.

Chlorosis in Grape Vines

Like many other plants, grape vines are susceptible to chlorosis, and symptoms of iron deficiency tend to be common on soils rich in limestone. In the wake of The Great French Wine Blight, when European *Vitis vinifera* were affected by *Phylloxera*, chlorosis became a greater problem in viticulture. To deal with the Phylloxera blight, *V. vinifera* was grafted onto rootstock based on American species of the *Vitis* genus, such as *Vitis riparia*, *Vitis rupestris*, *Vitis berlandieri*. However, many of these were less adapted to the lime-rich soils that were common in France's vineyards, in particular many of those that produced wines of top quality. Many grafted vines in lime-rich vineyards therefore showed signs of iron deficiency, and in France this specific form of chlorosis was termed *chlorose calcaire*. The problem was largely overcome by the selection of lime-resistant American vines as basis for hybrid vines used for rootstock material. However, since such rootstocks may be less than optimal in other respects, it is necessary for the viticulturalist to balance the need for chlorosis resistance against other viticultural needs. This is illustrated by one of the most common lime-resistant rootstocks, 41 B, which is a hybrid between *V. vinifera* cultivar Chasselas and *V. berlandieri*, which generally has a sufficient, but not extremely high, Phylloxera resistance.

Treatments

Specific nutrient deficiencies (often aggravated by high soil pH) may be corrected by supplemental feedings of iron, in the form of a chelate or sulphate, magnesium or nitrogen compounds in various combinations.

Fertilizer Burn

Fertilizer burn is defined as leaf scorch resulting from over-fertilisation, usually referring to excess nitrogen salts. Fertilizer burn is the result of desiccation of plant tissues due to osmotic stress, creating a state of hypertonicity. Fertilizers vary in their tendency to burn roughly in accordance with their salt index.

Treatment

Fertilizer burn can be remedied by soaking the soil with water to flush out excess salts. It can be prevented by applying only dilute or controlled-release fertilizer products.

Assimilation

Nitrogen Assimilation

Nitrogen assimilation is the formation of organic nitrogen compounds like amino acids from inorganic nitrogen compounds present in the environment. Organisms like plants, fungi and certain bacteria that cannot fix nitrogen gas (N_2) depend on the ability to assimilate nitrate or ammonia for their needs. Other organisms, like animals, depend entirely on organic nitrogen from their food.

Nitrogen Assimilation in Plants

Plants absorb nitrogen from the soil in the form of nitrate (NO_3^-) and ammonia (NH_3). In aerobic soils where nitrification can occur, nitrate is usually the predominant form of available nitrogen that is absorbed. However this need not always be the case as ammonia can predominate in grasslands and in flooded, anaerobic soils like rice paddies. Plant roots themselves can affect the abundance of various forms of nitrogen by changing the pH and secreting organic compounds or oxygen. This influences microbial activities like the inter-conversion of various nitrogen species, the release of ammonia from organic matter in the soil and the fixation of nitrogen by non-nodule-forming bacteria.

Ammonium ions are absorbed by the plant via ammonia transporters. Nitrate is taken up by several nitrate transporters that use a proton gradient to power the transport. Nitrogen is transported from the root to the shoot via the xylem in the form of nitrate, dissolved ammonia and amino acids. Usually (but not always) most of the nitrate reduction is carried out in the shoots while the roots reduce only a small fraction of the absorbed nitrate to ammonia. Ammonia (both absorbed and synthesized) is incorporated into amino acids via the glutamine synthetase-glutamate synthase (GS-GOGAT) pathway. While nearly all the ammonia in the root is usually incorporated into amino acids at the root itself, plants may transport significant amounts of ammonium ions in the xylem to be fixed in the shoots. This may help avoid the transport of organic compounds down to the roots just to carry the nitrogen back as amino acids.

Nitrate reduction is carried out in two steps. Nitrate is first reduced to nitrite (NO_2^-) in the cytosol by nitrate reductase using NADH or NADPH. Nitrite is then reduced to ammonia in the chloroplasts (plastids in roots) by a ferredoxin dependent nitrite reductase. In photosynthesizing tissues, it uses an isoform of ferredoxin (Fd1) that is reduced by PSI while in the root it uses a form of ferredoxin (Fd3) that has a less negative midpoint potential and can be reduced easily by NADPH. In non photosynthesizing tissues, NADPH is generated by glycolysis and the pentose phosphate pathway.

In the chloroplasts, glutamine synthetase incorporates this ammonia as the amide group of glutamine using glutamate as a substrate. Glutamate synthase (Fd-GOGAT and NADH-GOGAT) transfer the amide group onto an 2-oxoglutarate molecule producing two glutamates. Further transaminations are carried out make other amino acids (most commonly aspargine) from glutamine. While the enzyme glutamate dehydrogenase (GDH) does not play a direct role in the assimilation, it protects the mitochondrial functions during periods of high nitrogen metabolism and takes part in nitrogen remobilization.

pH and Ionic Balance During Nitrogen Assimilation

Every nitrate ion reduced to ammonia produces one OH^- ion. To maintain a pH balance, the plant must either excrete it into the surrounding medium or neutralize it with organic acids. This results in the medium around the plants roots becoming alkaline when they take up nitrate.

To maintain ionic balance, every NO_3^- taken into the root must be accompanied by either the uptake of a cation or the excretion of an anion. Plants like tomatoes take up metal ions like K^+, Na^+, Ca^{2+} and Mg^{2+} to exactly match every nitrate taken up and store these as the salts of organic acids like malate and oxalate. Other plants like the soyabean balance most of their NO_3^- intake with the excretion of OH^- or HCO_3^-.

Plants that reduce nitrates in the shoots and excrete alkali from their roots need to transport the alkali in an inert form from the roots to the shoots. To achieve this they synthesize malic acid in the leaves from neutral precursors like carbohydrates. The potassium ions brought to the leaves along with the nitrate in the xylem are then sent along with the malate to the roots via the phloem. In the roots, the malate is consumed. When malate is converted back to malic acid prior to use, an OH^- is released and excreted. ($RCOO^- + H_2O \rightarrow RCOOH + OH^-$) The potassium ions are then recirculated up the xylem with fresh nitrate. Thus the plants avoid having to absorb and store excess salts and also transport the OH^-.

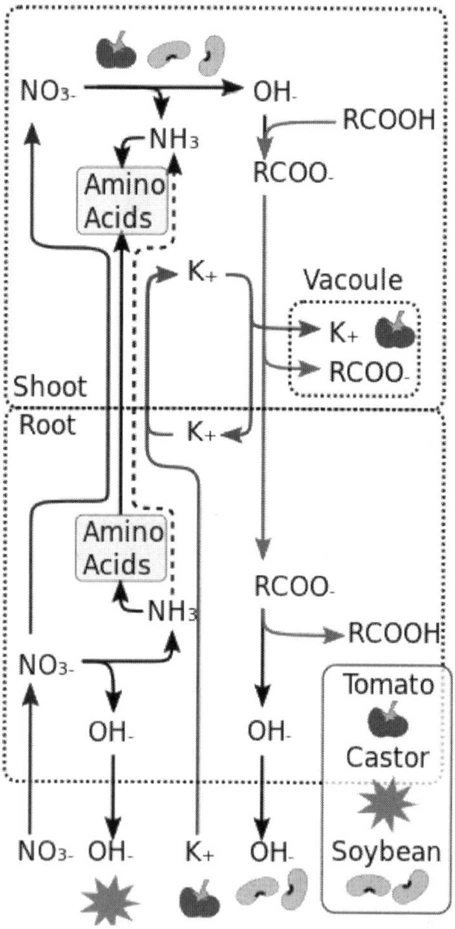

Figure: *Different plants use different pathways to different levels. Tomatoes take in a lot of K^+ and accumulate salts in their vacuoles, castor reduces nitrate in the roots to a large extent and excretes the resulting alkali. Soy bean plants moves a large amount of malate to the roots where they convert it to alkali while the potassium is recirculated.*

Plants like castor reduce a lot of nitrate in the root itself, and excrete the resulting base. Some of the base produced in the shoots is transported to the roots as salts of organic acids while a small amount of the carboxylates are just stored in the shoot itself.

Phosphate Solubilizing Bacteria

Phosphate solubilizing bacteria (PSB) are a group of beneficial bacteria capable of hydrolysing organic and inorganic phosphorus from insoluble compounds. P-solubilization ability of the microorganisms is considered to be one of the most important traits associated with plant phosphate nutrition. It is generally accepted that the mechanism of mineral phosphate solubilization by PSB strains is associated with the release of low molecular weight organic acids,through which their hydroxyl and carboxyl groups chelate the cations bound to phosphate, thereby converting it into soluble forms.In addition, some PSB produce phosphatase like phytase that hydrolyse organic forms of phosphate compounds efficiently. One or both types of PSB have been introduced to Agricultural community as phosphate Biofertilizer. Phosphorus (P) is one of the major essential macronutrients for plants and is applied to soil in the form of phosphate fertilizers. However, a large portion of soluble inorganic phosphate which is applied to the soil as chemical fertilizer is immobilized rapidly and becomes unavailable to plants. Currently, the main purpose in managing soil phosphorus is to optimize crop production and minimize P loss from soils. PSB have attracted the attention of agriculturists as soil inoculums to improve the plant growth and yield. When PSB used with rock phosphate, it can save about 50% of the crop requirement of phosphatic fertilizer. The use of PSB as inoculants increases the P uptake by plants. Simple inoculation of seeds with PSB gives crop yield responses equivalent to 30 kg P_2O_5 /ha or 50 percent of the need for phosphatic fertilizers. Currently, different strains of these bacteria has been identified for using in biofertilizer, of all, three new strains Pantoea agglomerans strain (P5), Microbacterium laevaniformans strain (P7) and Pseudomonas putida strain (P13) has been recently identified as the highly efficient insoluble phosphate solubilizer.

Sulphur Assimilation

Sulphur is an essential element for growth and physiological functioning of plants. However, its content strongly varies between plant species and it ranges from 0.1 to 6% of the plants' dry weight. Sulphates taken up by the roots are the major sulphur source for growth, though it has to be reduced to sulfide before it is further metabolized. Root plastids contain all sulphate reduction enzymes, but the reduction of sulphate to sulfide and its subsequent incorporation into cysteine predominantly takes place in the shoot, in the chloroplasts. Cysteine is the precursor or reduced sulphur donor of most other organic sulphur compounds in plants. The predominant proportion of the organic sulphur is present in the protein fraction (up to 70% of total sulphur), as cysteine and methionine (two amino acids) residues. Cysteine and methionine are highly significant in the structure, conformation and function of proteins. Plants contain a large variety of other organic sulphur compounds, as thiols (glutathione), sulfolipids and secondary sulphur compounds (alliins, glucosinolates, phytochelatins), which play an important role in physiology and protection against environmental stress and pests. Sulphur compounds are also of great importance for food quality and for the production of phyto-pharmaceutics. Sulphur deficiency will result in the loss of plant production, fitness and resistance to environmental stress and pests.

Sulphate Uptake by Plants

Sulphate is taken up by the roots that have high affinity. The maximal sulphate uptake rate is generally already reached at sulphate levels of 0.1 mM and lower. The uptake of sulphate by the roots and its transport to the shoot is strictly controlled and it appears to be one of the primary regulatory sites of sulphur assimilation. Sulphate is actively taken up across the plasma membrane of the root cells, subsequently loaded into the xylem vessels and transported to the shoot by the transpiration stream. The uptake and transport of sulphate is energy dependent (driven by a proton gradient generated by ATPases) through a proton/sulphate co-transport. In the shoot the sulphate is unloaded and transported to the chloroplasts where it is reduced. The remaining sulphate in plant tissue is predominantly present in the vacuole, since the concentration of sulphate in the cytoplasm is kept rather constant. Distinct sulphate transporter proteins mediate the uptake, transport and subcellular distribution of sulphate. According to their cellular and subcellular gene expression, and possible functioning the sulphate transporters gene family has been classified in up to 5 different groups. Some groups are expressed exclusively in the roots or shoots or expressed both in the roots and shoots. Group 1 are 'high affinity sulphate transporters', which are involved in the uptake of sulphate by the roots. Group 2 are vascular transporters and are 'low affinity sulphate transporters'. Group 3 is the so-called 'leaf group', however, still little is known about the characteristics of this group. Group 4 transporters are involved in the efflux of sulphate from the vacuoles, whereas the function of Group 5 sulphate transporters is not known yet, and likely function only as molybdate transporters. Regulation and expression of the majority of sulphate transporters are controlled by the sulphur nutritional status of the plants. Upon sulphate deprivation, the rapid decrease in root sulphate is regularly accompanied by a strongly enhanced expression of most sulphate transporter genes (up to 100-fold), accompanied by a substantially enhanced sulphate uptake capacity. It is not yet solved, whether sulphate itself or metabolic products of the sulphur assimilation (O-acetylserine, cysteine, glutathione) act as signals in the regulation of sulphate uptake by the root and its transport to the shoot, and in the expression of the sulphate transporters involved.

Sulphate Reduction in Plants

Even though root plastids contain all sulphate reduction enzymes, sulphate reduction predominantly takes place in the leaf chloroplasts. The reduction of sulphate to sulfide occurs in three steps. Sulphate needs to be activated to adenosine 5'-phosphosulphate (APS) prior to its reduction to sulphite. The activation of sulphate is catalyzed by ATP sulphurylase, which affinity for sulphate is rather low (Km approximately 1 mM) and the in situ sulphate concentration in the chloroplast is most likely one of the limiting/regulatory steps in sulphur reduction. Subsequently APS is reduced to sulphite, catalyzed by APS reductase with likely glutathione as reductant. The latter reaction is assumed to be one of the primary regulation points in the sulphate reduction, since the activity of APS reductase is the lowest of the enzymes of the sulphate reduction pathway and it has a fast turnover rate. Sulphite is with high affinity reduced by sulphite reductase to sulfide with ferredoxin as a reductant. The remaining sulphate in plant tissue is transferred into the vacuole. The remobilization and redistribution of the vacuolar sulphate reserves appear to be rather slow and sulphur-deficient plants may still contain detectable levels of sulphate.

Synthesis and Function of Sulphur Compounds in Plants

Cysteine: Sulfide is incorporated into cysteine, catalyzed by O-acetylserine (thiol)lyase, with O-acetylserine as substrate. The synthesis of O-acetylserine is catalyzed by serine acetyltransferase and together with O-acetylserine (thiol)lyase it is associated as enzyme complex named cysteine synthase. The formation of cysteine is the direct coupling step between sulphur (sulphur metabolism) and nitrogen assimilation in plants. This differs from the process in yeast, where sulfide must be incorporated first in homocysteine then converted in two steps to cysteine. Cysteine is sulphur donor for the synthesis of methionine, the major other sulphur-containing amino acid present in plants. This happens through the transsulphuration pathway and the methylation of homocysteine. Both cysteine and methionine are sulphur-containing amino acids and are of great significance in the structure, conformation and function of proteins and enzymes, but high levels of these amino acids may also be present in seed storage proteins. The thiol groups of the cysteine residues in proteins can be oxidized resulting in disulfide bridges with other cysteine side chains (and form cystine) and/or linkage of polypeptides. Disulfide bridges (disulfide bonds) make an important contribution to the structure of proteins. The thiol groups are also of great importance in substrate binding of enzymes, in metal-sulphur clusters in proteins (e.g. ferredoxins) and in regulatory proteins (e.g. thioredoxins).

Glutathione

Glutathione or its homologues, e.g. homoglutathione in Fabaceae; hydroxymethylglutathione in Poaceae are the major water-soluble non-protein thiol compounds present in plant tissue and account for 1-2% of the total sulphur. The content of glutathione in plant tissue ranges from 0.1 - 3 mM. Cysteine is the direct precursor for the synthesis of glutathione (and its homologues). First, γ-glutamylcysteine is synthesized from cysteine and glutamate catalyzed by gamma-glutamylcysteine synthetase. Second, glutathione is synthesized from γ-glutamylcysteine and glycine (in glutathione homologues, β-alanine or serine) catalyzed by glutathione synthetase. Both steps of the synthesis of glutathione are ATP dependent reactions. Glutathione is maintained in the reduced form by an NADPH-dependent glutathione reductase and the ratio of reduced glutathione (GSH) to oxidized glutathione (GSSG) generally exceeds a value of 7. Glutathione fulfils various roles in plant functioning. In sulphur metabolism it functions as reductant in the reduction of APS to sulphite. It is also the major transport form of reduced sulphur in plants. Roots likely largely depend for their reduced sulphur supply on shoot/root transfer of glutathione via the phloem, since the reduction of sulphur occurs predominantly in the chloroplast. Glutathione is directly involved in the reduction and assimilation of selenite into selenocysteine. Furthermore, glutathione is of great significance in the protection of plants against oxidative and environmental stress and it depresses/scavenges the formation of toxic reactive oxygen species, e.g.

superoxide, hydrogen peroxide and lipid hydroperoxides. Glutathione functions as reductant in the enzymatic detoxification of reactive oxygen species in the glutathione-ascorbate cycle and as thiol buffer in the protection of proteins via direct reaction with reactive oxygen species or by the formation of mixed disulfides. The potential of glutathione as protectant is related to the pool size of glutathione, its redox state (GSH/GSSG ratio) and the activity of glutathione reductase. Glutathione is the precursor for the synthesis of phytochelatins, which are synthesized enzymatically by a constitutive phytochelatin synthase. The number of γ-glutamyl-cysteine residues in the phytochelatins may range from 2 - 5, sometimes up to 11. Despite the fact that the phytochelatins form complexes which a few heavy metals, viz. cadmium, it is assumed that these compounds play a role in heavy metal homeostasis and detoxification by buffering of the cytoplasmatic concentration of essential heavy metals. Glutathione is also involved in the detoxification of xenobiotics, compounds without direct nutritional value or significance in metabolism, which at too high levels may negatively affect plant functioning. Xenobiotics may be detoxified in conjugation reactions with glutathione catalyzed by glutathione S-transferase, which activity is constitutive; different xenobiotics may induce distinct isoforms of the enzyme. Glutathione S-transferases have great significance in herbicide detoxification and tolerance in agriculture and their induction by herbicide antidotes ('safeners') is the decisive step for the induction of herbicide tolerance in many crop plants. Under natural conditions glutathione S-transferases are assumed to have significance in the detoxification of lipid hydroperoxides, in the conjugation of endogenous metabolites, hormones and DNA degradation products, and in the transport of flavonoids.

Sulfolipids

Sulfolipids are sulphur containing lipids. Sulfoquinovosyl diacylglycerols are the predominant sulfolipids present in plants. In leaves its content comprises up to 3 - 6% of the total sulphur present. This sulfolipid is present in plastid membranes and likely is involved in chloroplast functioning. The route of biosynthesis and physiological function of sulfoquinovosyl diacylglycerol is still under investigation. From recent studies it is evident that sulphite it the likely sulphur precursor for the formation of the sulfoquinovose group of this lipid.

Secondary Sulphur Compounds

Brassica species contain glucosinolates, which are sulphur-containing secondary compounds. Glucosinolates are composed of a β-thioglucose moiety, a sulfonated oxime and a side chain. The synthesis of glucosinolates starts with the oxidation of the parent amino acid to an aldoxime, followed by the addition of a thiol group (through conjugation with glutathione) to produce thiohydroximate. The transfer of a glucose and a sulphate moiety completes the formation of the glucosinolates. The physiological significance of glucosinolates is still ambiguous, though they are considered to function as sink compounds in situations of sulphur excess. Upon tissue disruption glucosinolates are enzymatically degraded by myrosinase and may yield a variety of biologically active products such as isothiocyanates, thiocyanates, nitriles and oxazolidine-2-thiones. The glucosinolate-myrosinase system is assumed to play a role in plant-herbivore and plant-pathogen interactions. Furthermore, glucosinolates are responsible for the flavour properties of Brassicaceae and recently have received attention in view of their potential anti-carcinogenic properties. Allium species contain γ-glutamylpeptides and alliins (S-alk(en)yl cysteine sulphoxide). The content of these sulphur-containing secondary compounds strongly depends on stage of development of the plant, temperature, water availability and the level of nitrogen and sulphur nutrition. In onion bulbs their content may account for up to 80% of the organic sulphur fraction. Less is known about the content of secondary sulphur compounds in the seedling stage of the plant. It is assumed that alliins are predominantly synthesized in the leaves, from where they are subsequently transferred to the attached bulb scale. The biosynthetic pathways of synthesis of γ-glutamylpeptides and alliins are still ambiguous. γ-Glutamylpeptides can be formed from cysteine (via γ-glutamylcysteine or glutathione) and can be metabolized into the corresponding alliins via oxidation and subsequent hydrolyzation by γ-glutamyl transpeptidases. However, other possible routes of the synthesis of γ-glutamylpeptides and alliins may not be excluded. Alliins and γ-glutamylpeptides are known to have therapeutic utility and might have potential value as phytopharmaceutics. The alliins and their breakdown products (e.g. allicin) are the flavour precursors for the odor and taste of species. Flavour is only released when plant cells are disrupted and the enzyme alliinase from the vacuole is able to degrade the alliins, yielding a wide variety of volatile and non-volatile sulphur-containing compounds. The physiological function of γ-glutamylpeptides and alliins is rather unclear.

Sulphur Metabolism in Plants and air Pollution

The rapid economic growth, industrialization and urbanization are associated with a strong increase in energy demand and emissions of air pollutants including sulphur dioxide and hydrogen sulfide, which may affect plant metabolism. Sulphur gases are potentially phytotoxic, however, they may also be metabolized and used as sulphur source and even be beneficial if the sulphur fertilization of the roots is not sufficient. Plant shoots form a sink for atmospheric sulphur gases, which can directly be taken up by the foliage (dry deposition). The foliar uptake of sulphur dioxide is generally directly dependent on the degree of opening of the stomates, since the internal resistance to this gas is low. Sulphur is highly soluble in the apoplastic water of the mesophyll, where it dissociates under formation of bisulfite and sulphite. Sulphite may directly enter the sulphur reduction pathway and be reduced to sulfide, incorporated into cysteine, and subsequently into other sulphur compounds. Sulphite may also be oxidized to sulphate, extra- and intracellularly by peroxidases or non-enzymatically catalyzed by metal ions or superoxide radicals and subsequently reduced and assimilated again. Excessive sulphate is transferred into the vacuole; enhanced foliar sulphate levels are characteristic for exposed plants. The foliar uptake of hydrogen sulfide appears to be directly dependent on the rate of its metabolism into cysteine and subsequently into other sulphur compounds. There is strong evidence that O-acetyl-serine (thiol)lyase is directly responsible for the active fixation of atmospheric hydrogen sulfide by plants. Plants are able to transfer from sulphate to foliar absorbed atmospheric sulphur as sulphur source and levels of 60 ppb or higher appear to be sufficient to cover the sulphur requirement of plants. There is an interaction between atmospheric and pedospheric sulphur utilization. For instance, hydrogen sulfide exposure may result in a decreased activity of APS reductase and a depressed sulphate uptake.

2

Biofertilizer

A biofertilizer (also *bio-fertilizer*) is a substance which contains living microorganisms which, when applied to seed, plant surfaces, or soil, colonizes the rhizosphere or the interior of the plant and promotes growth by increasing the supply or availability of primary nutrients to the host plant. Bio-fertilizers add nutrients through the natural processes of nitrogen fixation, solubilizing phosphorus, and stimulating plant growth through the synthesis of growth-promoting substances. Bio-fertilizers can be expected to reduce the use of chemical fertilizers and pesticides. The microorganisms in bio-fertilizers restore the soil's natural nutrient cycle and build soil organic matter. Through the use of bio-fertilizers, healthy plants can be grown, while enhancing the sustainability and the health of the soil. Since they play several roles, a preferred scientific term for such beneficial bacteria is "plant-growth promoting rhizobacteria" (PGPR). Therefore, they are extremely advantageous in enriching soil fertility and fulfilling plant nutrient requirements by supplying the organic nutrients through microorganism and their byproducts. Hence, bio-fertilizers do not contain any chemicals which are harmful to the living soil.

Bio-fertilizers provide eco-friendly organic agro-input and are more cost-effective than chemical fertilizers. Bio-fertilizers such as Rhizobium, Azotobacter, Azospirillum and blue green algae (BGA) have been in use a long time. Rhizobiuminoculant is used for leguminous crops. Azotobacter can be used with crops like wheat, maize, mustard, cotton, potato and other vegetable crops. Azospirillum inoculations are recommended mainly for sorghum, millets, maize, sugarcane and wheat. Blue green algae belonging to a general cyanobacteria genus, *Nostoc* or *Anabaena* or *Tolypothrix* or *Aulosira*, fix atmospheric nitrogen and are used as inoculations for paddy crop grown both under upland and low-land conditions. *Anabaena* in association with water fern Azolla contributes nitrogen up to 60 kg/ha/season and also enriches soils with organic matter.

Other types of bacteria, so-called phosphate-solubilizing bacteria, such as Pantoea agglomerans strain P5 or Pseudomonas putida strain P13, are able to solubilize the insoluble phosphate from organic and inorganic phosphate sources. In fact, due to immobilization of phosphate by mineral ions such as Fe, Al and Ca or organic acids, the rate of available phosphate (P_i) in soil is well below plant needs. In addition, chemical P_i fertilizers are also immobilized in the soil, immediately, so that less than 20 percent of added fertilizer is absorbed by plants. Therefore, reduction in P_i resources, on one hand, and environmental pollutions resulting from both production and applications of chemical P_i fertilizer, on the other hand, have already demanded the use of new generation of phosphate fertilizers globally known as phosphate-solubilizing bacteria or phosphate bio-fertilizers.

Benefits

A bio-fertilizer provides the following benefits:

1. Since a bio-fertilizer is technically living, it can symbiotically associate with plant roots. Involved microorganisms could readily and safely convert complex organic material in simple compounds, so that plants are easily taken up. Microorganism function is in long duration, causing improvement of the soil fertility. It maintains the natural habitat of the soil. It increases crop yield by 20-30%, replaces chemical nitrogen and phosphorus by 25%, and stimulates plant growth. It can also provide protection against drought and some soil-borne diseases.
2. Bio-fertilizers are cost-effective relative to chemical fertilizers. They have lower manufacturing costs, especially regarding nitrogen and phosphorus use.

Some important groups of Bio-fertilizers:

1. Azolla-Anabena symbiosis: Azolla is a small, eukaryotic, aquatic fern having global distribution. Prokaryotic blue green algae Anabena azolla resides in its leaves as a symbiont. Azolla is an alternative nitrogen source. This association has gained wide interest because of its potential use as an alternative to chemical fertilizers.

2. Rhizobium: Symbiotic nitrogen fixation by Rhizobium with legumes contribute substantially to total nitrogen fixation. Rhizobium inoculation is a well-known agronomic practice to ensure adequate nitrogen.

Photorespiration

Photorespiration (also known as the oxidative photosynthetic carbon cycle, or C2 photosynthesis) is a process in plant metabolism which attempts to ameliorate the consequences of a wasteful oxygenation reaction by the enzyme RuBisCO. The desired reaction is the addition of carbon dioxide to RuBP (carboxylation), a key step in the Calvin-Benson cycle, however approximately 25% of reactions by RuBisCO instead add oxygen to RuBP (oxygenation), producing a product that cannot be used within the Calvin-Benson cycle. This process reduces efficiency of photosynthesis, potentially reducing photosynthetic output by 25% in C_3 plants. Photorespiration involves a complex network of enzyme reactions that exchange metabolites between chloroplasts, leaf peroxisomes and mitochondria.

The oxygenation reaction of RuBisCO is a wasteful process because 3-Phosphoglycerate is created at a reduced rate and higher metabolic cost compared with RuBP carboxylase activity. While photorespiratory carbon cycling results in the formation of G3P eventually, there is still a net loss of carbon (around 25% of carbon fixed by photosynthesis is re-released as CO_2) and nitrogen, as ammonia. The ammonia must be detoxified at a substantial cost to the cell. Photorespiration also incurs a direct cost of 2ATP and one NAD(P)H.

While it is common to refer to the entire process as photorespiration, technically the term refers only to the metabolic network which acts to rescue the products of the oxygenation reaction (phosphoglycolate).

Photorespiratory Reactions

Addition of molecular oxygen to ribulose-1,5-bisphosphate produces 3-phosphoglycerate (PGA) and 2-phosphoglycolate (2PG, or PG). PGA is the normal product of carboxylation, and productively enters the Calvin cycle. Phosphoglycolate, however, inhibits certain enzymes involved in photosynthetic carbon fixation (hence is often said to be an 'inhibitor of photosynthesis'). It is also relatively difficult to recycle: in higher plants it is salvaged by a series of reactions in the peroxisome, mitochondria, and again in the peroxisome where it is converted into glycerate. Glycerate reenters the chloroplast and by the same transporter that exports glycolate. A cost of 1 ATP is associated with conversion to 3-phosphoglycerate (PGA) (Phosphorylation), within the chloroplast, which is then free to re-enter the Calvin cycle.

There are several costs associated with this metabolic pathway; one being the production of hydrogen peroxide in the peroxisome (associated with the conversion of glycolate to glyoxylate). Hydrogen peroxide is a dangerously strong oxidant which must be immediately broken down into water and oxygen by the enzyme catalase. The conversion of 2x 2Carbon glycine to 1 C3 serine in the mitochondria by the enzyme glycine-decarboxylase is a key step, which releases CO_2, NH_3, and reduces NAD to NADH. Thus, 1 CO_2 molecule is produced for every 3 molecules of O_2 (two deriving from the activity of RuBisCO, the third from peroxisomal oxidations). The assimilation of NH_3 occurs via the GS-GOGAT cycle, at a cost of one ATP and one NADPH.

Cyanobacteria have three possible pathways through which they can metabolise 2-phosphoglycolate. They are unable to grow if all three pathways are knocked out, despite having a carbon concentrating mechanism that should dramatically reduce the rate of photorespiration.

Substrate Specificity of RuBisCO

The oxidative photosynthetic carbon cycle reaction is catalyzed by RuBP oxygenase activity:

RuBP + O_2 → Phosphoglycolate + 3-phosphoglycerate + 2H+

During the catalysis by RuBisCO, an 'activated' intermediate is formed (an enediol intermediate) in the RuBisCO active site. This intermediate is able to react with either CO_2 or O_2. It has been demonstrated that the specific shape of the RuBisCO active site acts to encourage reactions with CO_2. Although there is a significant "failure" rate (~25% of reactions are oxygenation rather than carboxylation), this represents significant favouring of CO_2, when the relative abundance of the two gasses is taken into account: in the current atmosphere, O_2 is approximately 500 times more abundant, and in solution O_2 is 25X more abundant than CO_2. The ability of RuBisCO to specify between the two gasses is known as its selectivity factor (or Srel), and it varies between species (although there is not much variation within the vascular plants).

A suggested explanation into Rubisco's inability to discriminate completely between CO_2 and O_2 is that it is an evolutionary relic: The early atmosphere in which primitive plants originated contained very little oxygen, the early evolution of RuBisCO was not influenced by its ability to discriminate between O_2 and carbon dioxide.

Conditions Which Increase Photorespiration

Photorespiration rates are increased by:

Altered Substrate Availability: Lowered CO2 or Increased O2

Factors which influence this include the atmospheric abundance of the two gases, the supply of the gases to the site of fixation (i.e. in land plants: whether the stomata are open or closed), the length of thc liquid phase (how far these gases have to diffuse through water in order to reach the reaction site). For example, when the stomata are closed to prevent water loss during drought: this limits the CO_2 supply, while O_2 production within the leaf will continue. In algae (and plants which photosynthesise underwater) gases have to diffuse significant distances through water, which results in a decrease in the availability of CO_2 relative to O_2. It has been predicted that the increase in ambient CO_2 concentrations predicted over the next 100 years may reduce the rate of photorespiration in most plants by around 50%.

Increased Temperature

At higher temperatures RuBisCO is less able to discriminate between CO_2 and O_2. This is because the enediol intermediate is less stable. Increasing temperatures also reduce the solubility of CO_2, thus reducing the concentration of CO_2 relative to O_2 in the chloroplast.

High Light

Photorespiration may be used as a mechanism to dissipate excess energy at high irradiance levels.

Biological Adaptation to Minimize Photorespiration

Certain species of plants or algae have mechanisms to reduce uptake of molecular oxygen by RuBisCO. These are commonly referred to as Carbon Concentrating Mechanisms (CCMs), as they increase the concentration of CO_2 so that Rubisco is less likely to produce glycolate through reaction with O_2.

Biochemical Carbon Concentrating Mechanisms

Biochemical CCMs concentrate carbon dioxide in one temporal or spacial region, through metabolite exchange. C4 and CAM photosynthesis both use the enzyme Phosphoenolpyruvate carboxylase (PEPC) to add CO_2 to a 3-Carbon sugar. PEPC is faster than Rubisco, and more selective for CO_2.

C4

C4 plants capture carbon dioxide in their mesophyll cells (using an enzyme called Phosphoenolpyruvate carboxylase which catalyzes the combination of carbon dioxide with a compound called Phosphoenolpyruvate (PEP)), forming oxaloacetate. This oxaloacetate is then converted to malate and is transported into the bundle sheath cells (site of carbon dioxide fixation by RuBisCO) where oxygen concentration is low to avoid photorespiration. Here, carbon dioxide is removed from the malate and combined with RuBP by Rubisco in the usual way, and the Calvin cycle proceeds as normal. The CO_2 concentrations in the Bundle Sheath are approximately 10-20 fold higher than the concentration in the mesophyll cells. This ability to avoid photorespiration makes these plants more hardy than other plants in dry and hot environments, wherein stomata are closed and internal carbon dioxide levels are low. Under these conditions, photorespiration does occur in C4 plants, but at a much reduced level compared with C3 plants in the same conditions. C4 plants include sugar cane, corn (maize), and sorghum.

CAM (Crassulacean Acid Metabolism)

CAM plants, such as cacti and succulent plants, also use the enzyme PEP carboxylase to capture carbon dioxide, but only at night. Crassulacean acid metabolism allows plants to conduct most of their gas exchange in the cooler night-time air, sequestering carbon in 4 carbon sugars which can be released to the photosynthesizing cells during the day. This allows CAM plants to reduce water loss (transpiration) by maintaining closed stomata during the day. CAM plants usually display other water-saving characteristics, such as thick cuticles, stomata with small apertures, and typically lose around 1/3 of the amount of water per CO_2 fixed.

Algae

There have been some reports of algae operating a biochemical CCM: shuttling metabolites within single cells to concentrate CO2 in one area. This process is not fully understood and may not be biologically widespread.

Biophysical Carbon Concentrating Mechanisms

This type of carbon concentrating mechanism (CCM) relies on a contained compartment within the cell into which CO2 is shuttled, and where RuBisCO is highly expressed. In many species, biophysical CCMs are only induced under low carbon dioxide concentrations. Biophysical CCMs are more evolutionarily ancient than biochemical CCMs. There is some debate as to when biophysical CCMs first evolved, but it is likely to have been during a period of low carbon dioxide, after the great oxidation event (2.4 billion years ago). Low CO_2 periods occurred around 750, 650, and 320-270 million years ago.

Eukaryotic Algae

In nearly all species of eukaryotic algae (*Chloromonas* being one notable exception), upon induction of the CCM, ~95% of RuBisCO is densely packed into a single subcellular compartment: the pyrenoid. Carbon dioxide is concentrated in this compartment using a combination of CO_2 pumps, bicarbonate pumps, and carbonic anhydrases. The pyrenoid is not a membrane bound compartment, but is found within the chloroplast, often surrounded by a starch sheath (which is not thought to serve a function in the CCM).

Hornworts

Certain species of Hornwort are the only land plants which are known to have a biophysical CCM involving concentration of carbon dioxide within pyrenoids in their chloroplasts.

Cyanobacteria

Cyanobacterial CCMs are similar in principle to those found in Eukaryotic algae and Hornworts, but the compartment into which carbon dioxide is concentrated has several structural differences. Instead of the pyrenoid, cyanobacteria contain Carboxysomes, which have a protein shell, and linker proteins packing RuBisCO inside with a very regular structure. Cyanobacterial CCMs are much better understood than those found in Eukaryotes, partly due to the ease of genetic manipulation of prokaryotes.

Photorespiration may Serve Some Purpose

Reducing photorespiration may not result in increased growth rates for plants. Some research has suggested, for example, that photorespiration may be necessary for the assimilation of nitrate from soil. Thus, a reduction in photorespiration by genetic engineering or because of increasing atmospheric carbon dioxide due to fossil fuel burning may not benefit plants as has been proposed. Several physiological processes may be responsible for linking photorespiration and nitrogen assimilation: one is that photorespiration increases availability of NADH, which is required for the conversion of nitrate to nitrite; another link is that certain nitrite transporters also transport bicarbonate, and elevated CO_2 has been shown to suppress nitrite transport into chloroplasts.

Although photorespiration is greatly reduced in C4 species, it is still an essential pathway - mutants without functioning 2-phosphoglycolate metabolism cannot grow in normal conditions. One mutant was shown to rapidly accumulate glycolate.

Although the functions of photorespiration remain controversial, it is widely accepted that this pathway influences a wide range of processes from bioenergetics, photosystem II function, and carbon metabolism to nitrogen assimilation and respiration. The photorespiratory pathway is a major source of H_2O_2 in photosynthetic cells. Through H_2O_2 production and pyridine nucleotide interactions, photorespiration makes a key contribution to cellular redox homeostasis. In so doing, it influences multiple signalling pathways, in particular, those that govern plant hormonal responses controlling growth, environmental and defence responses, and programmed cell death.

Another theory postulates that it may function as a "safety valve", preventing the excess of reductive potential coming from an overreduced NADPH-pool from reacting with oxygen and producing free radicals, as these can damage the metabolic functions of the cell by subsequent oxidation of membrane lipids, proteins or nucleotides.

Methods

Fertigation

Fertigation is the application of fertilizers, soil amendments, or other water-soluble products through an irrigation system. Chemigation, a related and sometimes interchangeable term, is the application of chemicals through an irrigation system. Chemigation is considered to be a more restrictive and controlled process, due to the potential nature of the products being delivered - pesticides, herbicides, fungicides - to cause harm to humans, animals, and the environment. Therefore, chemigation is in general more regulated than fertigation.

Usage

Fertigation is used extensively in commercial agriculture and horticulture and is starting to be used in general landscape applications as dispenser units become more reliable and easy to use.

- Fertigation is used to spoon-feed additional nutrients or correct nutrient deficiencies detected in plant tissue analysis
- Usually practiced with high-value crops such as vegetables, turf, fruit trees, and ornamentals
- Injection during middle one-third or the middle one-half of the irrigation recommended for fertigation using micropropagation
- Water supply for fertigation kept separate from domestic water supply to avoid contamination
- Change of fertilizer programme during the growing season in order to adjust for fruit, flower, and root development.

Commonly Used Nutrients

- Most plant nutrients can be applied through irrigation systems
- Nitrogen is most commonly used nutrient
- Other nutrients include nitrate, ammonium, urea, phosphate, and potassium.

Determining Which Nutrient is Used

- A soil fertility analysis is used to determine which of the more stable nutrients should be used

Advantages

Benefits of fertigation over traditional broadcast or drop-fertilizing methods include:

- Increased nutrient absorption by plants
- Reduction in fertilizer and chemicals needed
- Reduced leaching to the water table
- Reduction in water usage due to the plant's resulting increased root mass's ability to trap and hold water
- Application of nutrients at the precise time they are needed and at the rate they are utilized

Disadvantages

- Concentration of solution decreases as fertilizer dissolves, leading to poor nutrient placement
- Results in pressure loss in main irrigation line
- Limited capacity
- Use of chemical fertilizers of low-sustainability, instead of organic fertilizers.

- Dependent on water supply's nonrestriction by drought rationing.

Controls

Because of the potential risk in contaminating the potable (drinking) water supply, a backflow prevention device is required for most fertigation systems. Backflow requirements vary greatly, so it is very important to understand the proper level of backflow prevention required by law. In the United States, the minimum backflow protection is usually determined by state regulation, but each city or town may increase the level of protection required.

Methods Used in Fertigation

- Drip irrigation, which reduces per water and nutrient application rates relative to sprinklers
- Sprinkler systems, which increase leaf and fruit quality.
- Other methods of application include lateral move, traveler gun, and solid set systems
- Continuous application - fertilizer is supplied at a constant rate
- Three-stage application - irrigation starts without fertilizers and then the later in process fertilizers are applied
- Proportional application - injection rate is proportional to water discharge rate
- Quantitative application - nutrient solution is applied in a calculated amount to each irrigation block

All systems should be placed on a raised and/or sealed platform, not in direct contact with the earth, and fitted with chemical spill trays.

In order to determine the injection rate for the particular fertilizer being used, one should use the formula:

Maximum injection rate = (5 x Q x L) / (f X 60)

where Q = irrigation pump discharge in litres per second, L = fertilizer tank volume in litres, and F = amount of fertilizer in grams.

System Design

Fertigation assists distribution of fertilizers for farmers. The simplest type of fertigation system consists of a tank with a pump, distribution pipes, capillaries, and dripper pen.

What Should be Considered

- Water quality
- Soil type
- Nutrient consumption (daily)
- Appropriate nutrient materials

Possible Strategies to be Used

- Injecting for short time-periods at the beginning, middle, and end of irrigation cycle
- Injecting during middle 50% of the irrigation cycle
- Continuous irrigation
- Postering index Imex

Fertilizer Tree

Fertilizer trees are used in agroforestry to improve the condition of soils used for farming. They capture nitrogen from the air and put it in the soil through their roots and falling leaves. They can also bring nutrients from deeper in the soil up to the surface for crops with roots that can't reach that depth. Fertilizer trees are further useful for preventing fertilizer erosion, and improving water usage for crops. Sesbania, Gliricidia, Tephrosia, and Faidherbia albida are known as fertilizer trees.

Use in Africa

Use of a fertilizer tree call the Faidherbia albida in Malawi and Zambia has resulted in a doubling or tripling of maize yields. As part of evergreen agriculture, use of fertilizer trees are proposed as a means to improve food security for Africa. Niger has more than 4.8 million hectares of predominantly Faidherbia agroforests, while Zambia has 300,000 hectares.

In Zambia and Malawi, where Faidherbia trees are most widely used, farmers plant the trees in a checkerboard pattern every 30 feet throughout the field.

Fertilizer trees are used to prevent the expansion of desert, and keep agriculture viable in arid lands in many African countries.

Green Manure

In agriculture, green manure refers to crops which have already been uprooted (and have often already been stuffed under the soil). The then dying plants are of a type of cover crop often grown primarily to add nutrients and organic matter to the soil. Typically, a green manure crop is grown for a specific period of time, and then plowed under and incorporated into the soil while green or shortly after flowering. Green manure crops are commonly associated with organic farming, and are considered essential for annual cropping systems that wish to be sustainable.

Functions

Green manures usually perform multiple functions, that include soil improvement and soil protection:

- Leguminous green manures such as clover and vetch contain nitrogen-fixing symbiotic bacteria in root nodules that fix atmospheric nitrogen in a form that plants can use. This performs the vital function of fertilization. If desired, animal manures may also be added.

Depending on the species of cover crop grown, the amount of nitrogen released into the soil lies between 40 and 200 pounds per acre. With green manure use, the amount of nitrogen that is available to the succeeding crop is usually in the range of 40-60% of the total amount of nitrogen that is contained within the green manure crop.

Average biomass yields and nitrogen yields of several legumes by crop:	*Biomass tons acre^{-1}*	*N lbs acre^{-1}*
Sweet clover	1.75	120
Berseem clover	1.10	70
Crimson clover	1.40	100
Hairy vetch	1.75	110

- Incorporation of cover crops into the soil allows the nutrients held within the green manure to be released and made available to the succeeding crops. This results immediately from an increase in abundance of soil microorganisms from the degradation of plant material that aid in the decomposition of this fresh material. This additional decomposition also allows for the re-incorporation of nutrients that are found in the soil in a particular form such as nitrogen (N), potassium (K), phosphorus (P), calcium (Ca), magnesium (Mg), and sulphur (S).
- Microbial activity from incorporation of cover crops into the soil leads to the formation of mycelium and viscous materials

which benefit the health of the soil by increasing its soil structure (i.e. by aggregation).

The increased percentage of organic matter (biomass) improves water infiltration and retention, aeration, and other soil characteristics. The soil is more easily turned or tilled than non-aggregated soil. Further aeration of the soil results from the ability of the root systems of many green manure crops to efficiently penetrate compact soils. The amount of humus found in the soil also increases with higher rates of decomposition, which is beneficial for the growth of the crop succeeding the green manure crop. Non-leguminous crops are primarily used to increase biomass.

- The root systems of some varieties of green manure grow deep in the soil and bring up nutrient resources unavailable to shallower-rooted crops.
- Common cover crop functions of weed suppression. Non-leguminous crops are primarily used (e.g. Buckwheat) The deep rooting properties of many green manure crops make them efficient at suppressing weeds.
- Some green manure crops, when allowed to flower, provide forage for pollinating insects. Green manure crops also often provide habitat for predatory beneficial insects, which allow for a reduction in the application of insecticides where cover crops are planted.
- Some green manure crops (e.g. Winter wheat and Winter rye) can also be used for grazing.
- Erosion control is often also taken into account when selecting which green manure cover crop to plant.
- Some green crops reduce plant insect pests and diseases. Verticillium wilt is especially reduced in potato plants.

Incorporation of green manures into a farming system can drastically reduce, if not eliminate, the need for additional products such as supplemental fertilizers and pesticides.

Limitations to consider in the use of green manure are time, energy, and resources (monetary and natural) required to successfully grow and utilize these cover crops. Consequently, it is important to choose green manure crops based on the growing region and annual precipitation amounts to ensure efficient growth and use of the cover crop(s).

Nutrient Creation

Green manure is broken down into plant nutrient components by heterotrophic bacteria that consumes organic matter. Warmth and moisture contribute to this process, similar to creating compost fertilizer. The plant matter releases large amounts of carbon dioxide and weak acids that react with insoluble soil minerals to release beneficial nutrients. Soils that are high in calcium minerals, for example, can be given green manure to generate a higher phosphate content in the soil, which in turn acts as a fertilizer.

The ratio of carbon to nitrogen in a plant is a crucial factor to consider, since it will impact the nutrient content of the soil and may starve a crop of nitrogen, if the incorrect plants are used to make green manure. The ratio of carbon to nitrogen will differ from species to species, and depending upon the age of the plant. The ratio is referred to as C:N. The value of N is always one, whereas the value of carbon or carbohydrates is expressed in a value of about 10 up to 90; the ratio must be less than 30:1 to prevent the manure bacteria from depleting existing nitrogen in the soil. Rhizobium are soil organisms that interact with green manure to retain atmospheric nitrogen in the soil. Legumes, such as beans, alfalfa, clover and lupines, have root systems rich in rhizobium, often making them the preferred source of green manure material.

Green Manure Crops

Winter green manure crops are oats and rye.

Other green manure crops:

- Alfalfa, which sends roots deep to bring nutrients to the surface.
- Buckwheat in temperate regions
- Cowpea
- Clover (e.g. annual sweet clover)
- Fava beans
- Fenugreek
- Lupin
- Millet
- Mustard
- Sesbania
- Sorghum
- Soybean

- Sudangrass
- Sunn hemp, a tropical legume
- Velvet bean (*Mucuna pruriens*), common in the southern US during the early part of the 20th century, before being replaced by soybeans, popular today in most tropical countries, especially in Central America, where it is the main green manure used in slash/mulch farming practices
- Vetch (*Vicia sativa*)
- Tyfon, a *Brassica* known for a strong tap root that breaks up heavy soils.
- Ferns of the genus *Azolla* have been used as a green manure in southeast Asia.

1. Jump up ^ This is one of many legumes that may be used as a green manure crop
2. ^ Jump up to: [a] [b] [c] [d] This is one of many non-legumes that may be used as a green manure crop
3. ^ Jump up to: [a] [b] [c] [d] [e] [f] This is one of many legumes that may be used as a green manure crop

History

The value of green manure was recognized by farmers in Ancient Greece, who ploughed broad bean plants into the soil. Chinese agricultural texts dating back hundreds of years refer to the importance of grasses and weeds in providing nutrients for farm soil. It was also known to early North American colonists arriving from Europe. Common colonial green manure crops were rye, buckwheat and oats. Traditionally, the incorporation of green manure into the soil is known as the fallow cycle of crop rotation, which was used to allow soils to recover.

Hoagland Solution

The Hoagland solution is a hydroponic nutrient solution that was developed by Hoagland and Arnon in 1933 and is one of the most popular solution compositions for growing plants (in the scientific world at least). The Hoagland solution provides every nutrient necessary for plant growth and is appropriate for the growth of a large variety of plant species. The solution described by Hoagland in 1933 has been modified several times (mainly to add iron chelates and the like), but the original concentrations for each element are shown below.

- N 210 ppm
- K 235 ppm
- Ca 200 ppm
- P 31 ppm
- S 64 ppm
- Mg 48 ppm
- B 0.5 ppm
- Fe 1 to 5 ppm
- Mn 0.5 ppm
- Zn 0.05 ppm
- Cu 0.02 ppm
- Mo 0.01 ppm

The Hoagland solution has a lot of N and K so it is very well suited for the development of large plants like Tomato and Bell Pepper. However, the solution is very good for the growth of plants with lower nutrient demands such as lettuce and aquatic plants with the further dilution of the preparation to 1/4 or 1/5. Hoagland solution must be made from 7stock.

1. Potassium nitrate(KNO_3)
2. Magnesium sulphate heptahydrate, $MgSO_4 \cdot 7H_2O$, and potassium dihydrogen phosphate (Potassium phosphate monobasic), KH_2PO_4
3. Iron EDTA or Iron chelate, Fe-EDTA
4. Boric Acid, H_3BO_3
5. Copper Sulphate, $CuSO_4$
6. Zinc sulphate heptahydrate, $ZnSO_4 \cdot 7H_2O$
7. Manganese chloride, $MnCl_2 \cdot 4H_2O$
8. Sodium molybdate, $Na_2MoO_4 \cdot 2H_2O$,
9. Calcium nitrate, $Ca(NO_3)_2 \cdot 4H_2O$. Last solution must be added as final.

Component	***Stock Solution***	***mL Stock Solution/1L***
Macronutrients		
2M KNO_3	202 g/L	2.5
2M KNO_3	202 g/L	2.5
2M $Ca(NO_3)_2 \cdot 4H_2O$	236 g/0.5L	2.5

Contd...

Component	Stock Solution	mL Stock Solution/1L
Iron (Sprint 138 iron chelate)	15 g/L	1.5
2M $MgSO_4 \cdot 7H_2O$	493 g/L	1
1M NH_4NO_3	80 g/L	1
	Micronutrients	
H_3BO_3	2.86 g/L	1
$MnCl_2 \cdot 4H_2O$	1.81 g/L	1
$ZnSO_4 \cdot 7H_2O$	0.22 g/L	1
$CuSO_4$	0.051 g/L	1
$H_3MoO_4 \cdot H_2O$ or	0.09 g/L	1
$Na_2MoO_4 \cdot {}_2O$	0.12 g/L	1
1M KH_2PO_4 (pH to 6.0)	136 g/L	0.5

1. Make up stock solutions and store in separate bottles with appropriate label.
2. Add each component to 800 mL deionized water then fill to 1 L.
3. After the solution is mixed, it is ready to water plants.

Hydroponic Dosers

Hydroponic dosers are used by indoor, hydroponic farmers to automate the task of dispensing pH solution and/or nutrient solution as necessary. When growing in hydroponics, the pH of the nutrient solution often drifts off target during use. The same is said for the amount of nutrients in the solution. These devices will automatically measure and adjust the solution as needed. By keeping TDS and pH levels in-range, plants grow efficiently, without health problems. Hydroponic dosers are generally specific for adjusting TDS or pH.

Living Mulch

In agriculture, a living mulch is a cover crop interplanted or undersown with a main crop, and intended to serve the functions of a mulch, such as weed suppression and regulation of soil temperature. Living mulches grow for a long time with the main crops, whereas cover crops are incorporated into the soil or killed with herbicides.

Other benefits of mulches are slowing the growth of weeds, and protecting soil from water and wind erosion. Some living mulches were found to increase populations of the natural enemies of crop pests. Legumes used as living mulches also provide nitrogen fixation, reducing the need for fertilizer.

Benefits

When cover crops are turned over into the soil, they contribute nutrients to the main crop so that less chemical fertilizer is required. The amount of the contribution depends on the biomass, which varies over time and depends on rainfall and other factors. The greater the biomass, the greater the nutrient turnover. Legume cover crops turn over nitrogen fixed from the atmosphere. Reports indicate that legumes in general have higher foliar nitrogen contents, from 20 to 45 mg g-1.

Bare soil resulting from intensive tillage can lead to soil erosion, nutrient losses, and offsite movement of pesticides. In addition, weeds can germinate and grow without competition. Living mulches can reduce water runoff and erosion, and protect waterways from pollution. Living mulches have also been shown to increase the population of organisms which are natural enemies of some crop pests.

Living mulches control weeds in two ways. When they are seeded before weed establishment, they suppress weeds by competition. In some situations, the allelopathic properties of living mulches can be used to control weeds. For example, the allelopathic properties of winter rye (*Secale cereale*), ryegrasses (*Lolium spp*), and subterrain clover (*Trifolium subterraneum*) can be used to control weeds in sweet corn (*Zea mays* var "rugosa") and snap beans (*Phaseolus vulgaris*).

Populations of ground-dwelling predators were greater in a corn and soybean rotation with alfalfa and kura clover living mulches than without a living mulch. This situation was due in part to a change in the composition of vegetation in the agricultural system.

Drawbacks

Unfortunately, living mulches compete for nutrients and water with the main crop, and this can reduce yields. For example, Elkins et al. (1983) examined the use of tall fescue (*Festuca arundinacea*), smooth bromegrass (*Bromus inermis*), and orchargrass (*Dactylis glomerata*) as living mulches. They found that herbicides killed 50% to 70% of the mulches but corn yield was reduced 5% to 10% at the end of the harvest.

Although leguminous cover crops have large biomass production and turnover, they are not likely to increase soil organic matter. This is because legumes used as living mulches have greater N contents and a low C to N ratio. So when legume residue decomposes, soil microbes have sufficient N available to enhance their breakdown of organic materials in the soil.

Types

In the tropics, it is common to seed tree crops with living mulches such as legume covers in oil palm plantations, coconut and rubber (Watson 1989). In Mexico, legumes used traditionally as living mulches were tested as nematode and weed suppressors. The mulches included velvetbean (*Mucuna pruriens*) jackbean (*Canavalia ensiformis*), jumbiebean (*Leucaena leucocephala*) and wild tamarindo (*Lysiloma latisiliquum*). Aqueous extract of Velvetbean reduced the gall index of *Meloidogyne incognita* in the roots of tomato, but suppressed tomato rooting as well. In addition, Velvetbean suppressed the radical growth of the local weeds Alegria (*Amaranthus hypochondriacus*) by 66% and Barnyardgrass (*Echinochloa crus-galli*) by 26.5%.

Nicholson and Wien (1983) suggested the use of short turfgrasses and clovers as living mulches to improve the resistance soil compaction. These authors established Smooth Meadow-grass (*Poa pratensis*) and white clover (*Trifolium repens*) as living mulches since they did not caused reduction of yield corn (the accompanying main crop).

In one study, chewing fescue or red fescue (*Festuca rubra*) and ladino clover (*Trifolium repens*) were effective living mulches for controlling weed growth. Unfortunately, these cover crops also competed with corn for water which was particularly problematic during a dry period. The possibility of using ladino clover (*Trifolium repens*) as a potential living mulch was also mentioned; however, this clover was difficult to kill with herbicides in winter.

Management

Because they compete with the main crop, living mulches may eventually need to be mechanically or chemically killed. It is important to judiciously select the appropriate herbicide rate for burning down a living mulch. In 1989, Echtenkamp and Moomaw found that herbicide rates were inadequate to suppress all the living mulches. Therefore, the mulches competed with the main crop for resources. In some cases, the clover could not be killed at the first herbicide application, so a second application was needed. For another treatment, rates that were so high that they caused the cover crop to be killed too rapidly, so that broadleaf weeds invaded the corn. This study suggested that the timing and dosage of herbicide should be carefully considered.

Living mulches were tested in a no-till corn-production systems with two methods for establishing grass and legume living mulches (grass and legume) between corn rows. In 1985, there was no difference between drilling and broadcasting seeds by hand in that study. However in 1986, drilling resulted in higher populations (97 plants m-2) than broadcasting (64 plants m-2), likely because of precipitation levels. Precipitation should be considered because farmers have no control over it.

Beard (1973) recommended chewing fescue (red fescue) (*Festuca rubra* var "commutata" var "shadow FESRU") as a good living mulch because it adapts to the shady conditions under corn and soybean. This grass is also well adapted to dry and poor soils.

Plant Nutrition

Legume cover crops have important positive effects on the nutrient cycling of tree crops. Leguminous living mulches work in three ways:

- Fixing the atmospheric N2 that is important for the main crop,
- Recycling soil nutrients, and
- Enhancing soil nutrient availability for the main crop.

Lehmann et al. (2000) measured the above ground biomass accumulation of *Pueraria phaseoloides,* which is a living mulch used in tropical tree crops. They found that *Pueraria* accumulated 838 mg of dm (dry matter) ha-1 as compared with 4.4 mg dm ha^{-1} for *Theobroma grandiflorum*, and 1.4 mg dm ha-1 for *Bactris gasipaes*. These latter two species are native cultivated species from the Amazon.

Control of Erosion

Vegetative cover as living mulches protect soil against wind and water erosion. Plants should form a mantle or thick mulch that protect soil from detachment. Living mulches intercept raindrops and reduce runoff. The protection that such vegetation provides against wind is influenced mainly by the amount of biomass that covers the ground (differs with each spp), plant geometry and row orientation.

In one experiment, water runoff and soil loss on a 14% slope was compared for rototilled (RT), no-till with corn stover mulch (NTCMS), no-till in CSM+ birdsfoot trefoil living mulch (NT-BFT) and no till in CSM and crownvetch living mulch (NT-CV). The results indicated that the water runoff was 6,350 L ha-1 for NT-BFT, 6,350 L ha-1 NO-CSM, 5,925 L ha-1 for NT-CV, and 145,000 L ha-1 for RT. The soil loss for the RT was 14.22 t ha-1 while with the other treatments it was less than 0.5 tons ha-1. The least soil loss was obtained with NT-CV 0.02 tons ha-1. The reduction of water runoff and erosion is one of the greatest advantages of having a cover crop. Soil can be easily eroded with no vegetative ground cover or plant residue. Ideally soil erosion should be less than 4 to 5 tons/ha/year.

Nutrient Budgeting

Nutrient budgeting is used in agriculture. The process involves balancing nutrients coming into the farming system with those leaving. The aim is to prevent pollution events and save costs by precisely matching the nutrient requirements of the crop with application of organic and inorganic fertilizers.

Nutrients entering the farming system from:

- Residual nutrients in the soil and organic matter remaining in the soil from previous crops
- Green manure
- Animal manures and slurries
- Inorganic fertilizer application

are balanced with nutrients leaving the farming system in crops.

Nutrient Management

Nutrient management is a system used by farmers to manage the amount, form, placement, and timing of the application of nutrients (whether as manure, commercial fertilizer, or other form of nutrients) to plants. The purpose is to supply plant nutrients for optimum forage and crop yields, to minimize nonpoint source pollution (runoff of pollutants to surface water) and contamination of groundwater, and to maintain and/or improve the condition of soil.

Nutrient Management Plan

A nutrient management plan is a set of conservation practices designed to use fertilizer and/or manure to effectively provide needed crop nutrients while protecting against the potential adverse impacts of manure, erosion and organic byproducts on water quality. When such a plan is designed for animal feeding operations (AFO) it may be termed a "manure management plan." The plans may address:

- soil testing
- Plant tissue testing
- manure testing
- erosion control practices

- According to the soil pH the recommend dose may be varying.
- timing of fertilizer/manure application.

In the United States some regulatory agencies recommend or require that farms implement these plans, in order to prevent water pollution. The U.S. Natural Resources Conservation Service (NRCS) has published guidance documents on preparing a comprehensive nutrient management plan (CNMP) for AFOs.

Organic Fertilizer

Organic fertilizers are fertilizers derived from animal or vegetable matter. (e.g. compost, manure). Naturally occcuring organic fertilizers include manure, slurry, worm castings, peat, seaweed, humic acid, and guano. Sewage sludge use in organic agricultural operations in the U.S. has been extremely limited and rare due to USDA prohibition of the practice (due to toxic metal accumulation, among other factors).

Processed organic fertilizers include compost, bloodmeal, bone meal, humic acid, amino acids, and seaweed extracts. Other examples are natural enzyme-digested proteins, fish meal, and feather meal. Decomposing crop residue (green manure) from prior years is another source of fertility.

Discussion of the Term 'Organic'

There used to be a distinction between the term "organic" and the term "pesticide-free". Organic simply dealt with the use of fertilizer types. Once the term "organic" became regulated, many other factors were added. "Pesticide-free" is not at all related to fertilization (plant nutrition), but has become a legal inclusion. In scientific terms, organic compounds are any chemical compounds that include carbon, regardless of their source.

Likewise, in scientific terms, a fish emulsion can be a good organic fertilizer, but in some jurisdictions{{like Oregon | date=November 2010}} fish emulsion must be certified "dolphin-safe" to be considered "organic".

Natural Sourcing

Animal sourced urea and urea-formaldehyde from urine are suitable for organic agriculture; however, synthetically produced urea is not. The common thread that can be seen through these examples is that *organic* agriculture attempts to define itself through minimal processing (e.g., via chemical energy such as petroleum), as well as being naturally occurring or via natural biological processes such as composting.

Cover crops are also grown to enrich soil as a green manure through nitrogen fixation from the atmosphere; as well as phosphorus (through nutrient mobilization) content of soils.

Powdered limestone, mined rock phosphate, and Chilean saltpeter are *inorganic* chemicals in the technical (organic chemistry) sense of the word but are considered suitable for organic agriculture in *limited amounts*.

Fertilizer trees aid organic farming by bringing nutrients from the depths of the soil, and by assisting in the regulation of water usage.

Advantages

Although the density of nutrients in organic material is comparatively modest, they have many advantages. The majority of nitrogen-supplying organic fertilizers contain insoluble nitrogen and act as a slow-release fertilizer. By their nature, organic fertilizers increase physical and biological nutrient storage mechanisms in soils, mitigating risks of over-fertilization. Organic fertilizer nutrient content, solubility, and nutrient release rates are typically much lower than mineral (inorganic) fertilizers. A University of North Carolina study found that potential mineralizable nitrogen (PMN) in the soil was 182–285% higher in organic mulched systems than in the synthetics control.

Organic fertilizers also re-emphasize the role of humus and other organic components of soil, which are believed to play several important roles:

- Mobilizing existing soil nutrients, so that good growth is achieved with lower nutrient densities while wasting less
- Releasing nutrients at a slower, more consistent rate, helping to avoid a boom-and-bust pattern
- Helping to retain soil moisture, reducing the stress due to temporary moisture stress
- Improving the soil structure
- Helping to prevent topsoil erosion (responsible for desertfication and the Dust bowl

Organic fertilizers also have the advantage of avoiding certain problems associated with the regular heavy use of artificial fertilizers:

- The necessity of reapplying artificial fertilizers regularly (and perhaps in increasing quantities) to maintain fertility

- Extensive runoff of soluble nitrogen and phosphorus, leading to eutrophication of bodies of water (which causes fish kills)
- Costs are lower for if fertilizer is locally available

Inconveniences

Organic fertilizers have the following inconveniences:

- As a dilute source of nutrients when compared to inorganic fertilizers, transporting large amount of fertilizer incurs higher costs, especially with slurry and manure.
- The composition of organic fertilizers tends to be more complex and variable than a standardized inorganic product.
- *Improperly* processed organic fertilizers may contain pathogens from plant or animal matter that are harmful to humans or plants. However, proper composting should remove them.
- More labour is needed to compost organic fertilizer, increasing labour costs. Some of this cost is offset by reduced cash purchase.
- More applications of organic fertilizer are needed to apply sufficient nutrients to the soil. The multiple passes of fertilizing equipment cause soil compaction, require more labour and use up more fuel.

Conventional Farming Application

In non-organic farming, a compromise between the use of artificial and organic fertilizers is common, often using inorganic fertilizers supplemented with the application of organics that are readily available such as the return of crop residues or the application of manure.

Plant Tissue Test

The nutrient content of a plant can be assessed by testing a sample of tissue from that plant. These tests are important in agriculture since fertilizer application can be fine-tuned if the plants nutrient status is known. Nitrogen most commonly limits plant growth and is the most managed nutrient.

Comparison to Soil Tests

Plant tissue tests are often used in combination with soil tests. Soil tests measure the concentrations of nutrients in the soil that are potentially available for uptake by plant roots.

Although soil testing is widely used, it does not always accurately estimate the nutrient status of the crop. Even if soils have the recommended level of nutrients, the crop can still be nitrogen deficient if it has a nutrient uptake problem. Only tissue tests tell growers the actual nitrogen status of the crop.

Another drawback of soil nutrient tests is that they do not take account of factors such as soil structure or biological activity, which affect the rate at which nutrients leave the soil. These factors mean that the concentration of nutrients in the soil can rapidly deviate from the results of a soil test.

Another reason why soil tests do not always guarantee the best management of nutrient treatments is that cultivars vary in their ability to scavenge soil nutrients. Some cultivars can produce their optimum yield when grown in soil with less than the recommended levels of nutrients. Directly testing the nitrogen status of the crop helps growers adjust fertilizer applications to the specific needs of that crop.

Most Useful Times

Tissue tests are almost always useful, since they provides additional information about the physiology of the crop. Tissue tests are especially useful in certain situations;

- For monitoring the nitrogen status of the crop throughout the growing season. Soil tests are commonly performed before planting.
- In highly controlled environments, such as hydroponic production in greenhouses, crops require a constant feed of nutrients in their water supply. Even a transient lack of nutrients can reduce yields. In these controlled environments, soil testing is unlikely to be sufficient to manage crop nitrogen status. Soil testing is more suitable when growing crops in slow-release composts and manures.
- When there is a risk that the nutrient applications are toxic to the crop, such as during the application of poultry litter that contains micro nutrients such as copper.
- To guarantee that nitrogen levels in the crop do not exceed a certain limit. High concentrations of nitrates has implications to human health because nitrates can be converted into nitrites in the human digestive tract. Nitrites can react with other compounds in the gut to form nitrosamines, which appear to be carcinogenic. Crops contain high concentrations of Nitrate when excess fertilizer is used. This is an issue in crops with high levels of nitrates, such as spinach and lettuce.

Disadvantages of Traditional Tests

Traditional tissue tests are destructive tests where a sample is sent to a laboratory for analysis. Any laboratory test (soil or tissue test) performed by a commercial company will cost the grower a fee. Laboratory tests take at least a week to complete, usually 2 weeks. It takes time to dry the samples, send them to the lab, complete the lab-tests, and then return the results to the grower. This means the results may not be received by the grower until after the ideal time to take action. Nitrogen tissue tests that can be performed quickly in the field make tissue testing much more useful.

Another issue with laboratory tissue tests is that the results are often difficult to interpret.

Non-destructive Tissue Tests

Non-destructive tissue tests have advantages over traditional destructive tests. Non-destructive tissue tests can be performed easily in the field, and provide results much faster than laboratory tests.

To non-destructively assess nitrogen content, one can assess the chlorophyll content. Nitrogen content is linked to chlorophyll content because a molecule of chlorophyll contains four nitrogen atoms.

Chlorophyll Content Metres

Nitrogen deficiency can be detected with a chlorophyll content metre. Many studies have used chlorophyll content metres to predict N-content of leaves, and generally a good correlation is obtained. Researchers at the University of Nebraska concluded that a producer's ability to manage nitrogen is enhanced by using a Chlorophyll Metre in combination with the researcher's new technique to fine-tune N-management during the growing season.

Peer-reviewed studies have concluded the CCM-200plus Chlorophyll Metre accurately measures leaf nitrogen content. For instance, researchers at the Universidad Politécnica de Madrid used a CCM-200plus to measure a Chlorophyll Content Index (CCI), and concluded that leaf chlorophyll content performed well as an indicator of nutritional status in sugar beet. The CCI was responsive to differences in nitrogen availability and was able to discriminate between treatments. The CCI parameter was concluded to be quick, easy and non-invasive.

Researchers at the University of Vermont assessed the ability of the CCM-200plus to estimate Chlorophyll Content Index (CCI) and nitrogen content. The N-content was laboratory analysed, and was found to have a significantly linear relationship with CCI ($P < 0.001$). The CCM-200plus was concluded to be an effective tool for estimating relative chlorophyll and nitrogen content.

Chlorophyll Fluorometry

An instrumental method of detecting nitrogen deficiency is with a chlorophyll fluorometer. Nitrogen deficiency tests based on the fluorescence emission of leaves have been developed. Certain spectrums of laser-induced fluorescence are used, since this particular spectrum is able to distinguish nitrogen and sulphur deficiency. Only N-deficiency significantly increases the fluorescence induced by 360nm- and 440nm-laser pulses. A test protocol based on this fluorescence spectrum has been developed and is incorporated into the some recent chlorophyll fluorometers, such as ADC's OS5p (in which it is called the FRFex360/FRFex440 test). This spectrum of laser-induced fluorescence changes prior to or simultaneously with plant growth inhibition, allowing this fluorescence ratio to quickly detect N-deficiency. Another test protocol based on fluorescence is the OJIP test This method analyses the increase in fluorescence emitted from dark-adapted leaves when they are illuminated. The rise in fluorescence during the first second of illumination follows a curve with intermediate peaks, called the O, J, I, and P steps. In addition, the K step appears during specific types of stress, such as N-deficiency. Research has shown the K step is able to measure N-stress.

Miscellaneous

Fertility (Soil)

Fertile soil has the following properties:

- It is rich in nutrients necessary for basic plant nutrition, including nitrogen,
- It contains sufficient minerals (trace elements) for plant nutrition, including boron, chlorine, cobalt, copper, iron, manganese, magnesium, molybdenum, sulphur, and zinc.
- It contains soil organic matter that improves soil structure and soil moisture retention.
- Soil pH is in the range 6.0 to 6.8 for most plants but some prefer acid or alkaline conditions.
- Good soil structure, creating well drained soil, but some soils are wetter (as for producing rice) or drier (as for producing plants susceptible to fungi or rot) such as agave.

- A range of microorganisms that support plant growth.
- It often contains large amounts of topsoil.

In lands used for agriculture and other human activities, fertile soil typically arises from the use of soil conservation practices. Basically the ability of a soil to supply plant nutrients.

Soil Fertilization

Bioavailable nitrogen is the element in soil that is most often lacking. Phosphorus and potassium are also needed in substantial amounts. For this reason these three elements are always identified on a commercial fertilizer analysis. For example a 10-10-15 fertilizer has 10 percent nitrogen, 10 percent (P_2O_5) available phosphorus and 15 percent (K_2O) water soluble potassium. Sulphur is the fourth element that may be identified in a commercial analysis—e.g. 21-0-0-24 which would contain 21% nitrogen and 24% sulphate.

Inorganic fertilizers are generally less expensive and have higher concentrations of nutrients than organic fertilizers. Some have criticized the use of inorganic fertilizers, claiming that the water-soluble nitrogen doesn't provide for the long-term needs of the plant and creates water pollution. Slow-release fertilizers may reduce leaching loss of nutrients and may make the nutrients that they provide available over a longer period of time.

Soil fertility is a complex process that involves the constant cycling of nutrients between organic and inorganic forms. As plant material and animal wastes decompose they release nutrients to the soil solution. Those nutrients may then undergo further transformations which may be aided or enabled by soil micro-organisms. Natural processes such as lightning strikes may fix atmospheric nitrogen by converting it to (NO_2). Denitrification may occur under anaerobic conditions (flooding) in the presence of denitrifying bacteria. The cations, primarily phosphate and potash, as well as many micronutrients are held in relatively strong bonds with the negatively charged portions of the soil in a process known as Cation Exchange Capacity.

In 2008 the cost of phosphorus as fertilizer more than doubled, while the price of rock phosphate as base commodity rose eight-fold. Recently the term peak phosphorus has been coined, due to the limited occurrence of rock phosphate in the world.

Light and CO_2 Limitations

Photosynthesis is the process whereby plants use light energy to drive chemical reactions which convert CO_2 into sugars. As such, all plants require access to both light and carbon dioxide to produce energy, grow and reproduce.

While typically limited by nitrogen, phosphorus and potassium, low levels of carbon dioxide can also act as a limiting factor on plant growth. Peer-reviewed and published scientific studies have shown that increasing CO_2 is highly effective at promoting plant growth up to levels over 300 ppm. Further increases in CO_2 can, to a very small degree, continue to increase net photosynthetic output.

Since higher levels of CO_2 have only a minimal impact on photosynthetic output at present levels (presently around 400 ppm and increasing), we should not consider plant growth to be limited by carbon dioxide. Other biochemical limitations, such as soil organic content, nitrogen in the soil, phosphorus and potassium, are far more often in short supply. As such, neither commercial nor scientific communities look to air fertilization as an effective or economic method of increasing production in agriculture or natural ecosystems. Furthermore, since microbial decomposition occurs faster under warmer temperatures, higher levels of CO_2 (which is one of the causes of unusually fast climate change) should be expected to increase the rate at which nutrients are leached out of soils and may have a negative impact on soil fertility.

Soil Depletion

Soil depletion occurs when the components which contribute to fertility are removed and not replaced, and the conditions which support soil's fertility are not maintained. This leads to poor crop yields. In agriculture, depletion can be due to excessively intense cultivation and inadequate soil management.

Karl Marx wrote of the role of capitalism in soil depletion. In Capital, Volume I, he wrote:

...all progress in capitalistic agriculture is a progress in the art, not only of robbing the labourer, but of robbing the soil; all progress in increasing the fertility of the soil for a given time, is a progress towards ruining the lasting sources of that fertility. The more a country starts its development on the foundation of modern industry, like the United States, for example, the more rapid is this process of destruction. Capitalist production, therefore, develops technology, and the combining together of various processes into a social whole, only by sapping the original sources of all wealth — the soil and the labourer.

Soil fertility can be severely challenged when land use changes rapidly. For example in Colonial New England, colonists made a number of decisions that depleted the soils, including: allowing herd animals to wander freely, not replenishing soils with manure, and a sequence of events that led to erosion. Writes William Cronon, "...the long-term effect was to put those soils in jeopardy. The removal of the forest, the increase in destructive floods, the soil compaction and close-cropping wrought by grazing animals, plowing—all served to increase erosion."

One of the most widespread occurrences of soil depletion as of 2008 is in tropical zones where nutrient content of soils is low. The combined effects of growing population densities, large-scale industrial logging, slash-and-burn agriculture and ranching, and other factors, have in some places depleted soils through rapid and almost total nutrient removal.

Topsoil depletion occurs when the nutrient-rich organic topsoil, which takes hundreds to thousands of years to build up under natural conditions, is eroded or depleted of its original organic material. Historically, many past civilizations' collapses can be attributed to the depletion of the topsoil. Since the beginning of agricultural production in the Great Plains of North America in the 1880s, about one-half of its topsoil has disappeared.

Depletion may occur through a variety of other effects, including overtillage (which damages soil structure), underuse of nutrient inputs which leads to mining of the soil nutrient bank, and salinization of soil.

Soil Organisms

The cycles that permit nutrients to flow from soil to i plant are all interdependent and proceed only with the help of living organisms that constitute the living community of the soil. Soil micro-organisms are the essential link between soil mineral constituents and plant growth. AnimaJs and people are also part of this community. Unless their wastes are returned to the soil for the benefit of these organisms that live there, the whole life support. ing process is undermined.

Soil organisms, from large-sized earthworms to , extremely minute bacteria and fungi, perform a vast array of fertility maintenance task. Most of the soil organisms ! are useful for the plants but some are harmful and either ! cause diseases or exist as pests. However, key to soil fertility lies in promoting population of soil organisms as they form a vital part of the soil ecosystem. Excess use of chemicals in the form of fertilizers and pesticides sometimes inhibit the growth of useful organism and sabotages the soil ecosystem temporarily. Therefore, this fact must be kept in mind while handling the soil.

The creatures inhabiting the soil represent an entire universe of life forms and interactions; their number can be counted in millions in every gram of healthy soil. The organisms in any ecosystem are classified on the basis of their source of energy into three broad groups, namely producers, consumers and decomposers. Producers are almost always green plants. Consumers are those which feed upon products synthesized by producers or on producers themselves. Decomposers serve the critical role of bringing nutrients full circle from consumers back to producers.

Consumers are grouped into two categories namely (l) primary consumers and (2) secondary and tertiary consumers.

Primary Consumers

As soon as a leaf or a stalk or a piece of bark drops on the ground, it is subjected to a coordinated attack by microflora (micro-organisms belonging to plant kingdom ego bacteria) and other organisms that live on dead and decaying plant tissues. If a little moisture is present, bacteria and fungi initiate the attack. They are then joined by mites, snails, beetles, millipedes, woodlike collembola, earthworms and enchytraeid worms. These animals chew or make holes in the tissues, opening it up to a more rapid attack by the microflora. Together with micro- flora, they utilize the energy stored in the plant residues and are termed primary consumers.

While the action of microflora is mostly chemical, that of the fauna is both physical and chemical. The animals chew the plant parts and move them from one place to another on the soil surface and even into the soil. Earthworms incorporate the plant residues into the mineral soil by literally eating their way through the soil. Larger animals such as gophers, moles, prairie dogs and rats also burrow into the soil and bring about consider- able soil mixing and granulation. Organisms such as collembola, mites and enchytraeids worms, edge their way into cracks and crevices in the soil.

Secondary and Tertiary Consumers

The primary consumers are themselves food sources for predators and parasites existing in the soil. Predators and parasites are the terms used for those organisms, which either eat or derive their food from other living organisms. These are 'secondary consumers' and include carnivores (animal consumers) and microphytic feeders which consume small plants (bacteria, fungi, algae and lichens). Examples of carnivores are centipedes which i consume small insects, collembola, spiders, nematodes, snails, and European mole which feeds primarily on earthworms. Mites, springtails, termites and protozoa utilize microflora as their source of food.

Moving further up the food chain the secondary consumers are prey for still other carnivores called 'tertiary consumers'. For example ants consume centipedes, spiders; mites and scorpions which themselves prey on I primary or other secondary consumers. Again the micro-flora are intimately involved in the decomposition of organic material associated with the fauna, they are active within the digestive tract of some of the animals. They also attack the finely shredded organic material in animal faeces and later decompose the bodies of dead animals.

Kinds of Soil Organisms

Grouping of soil organisms discussed above is based on their feeding habits or energy requirement. The terms 'micro-organisms and 'macro-organisms' are also frequently used to group soil organisms according to their size. Organisms which can be seen by unaided eye are called macro-organisms and the ones which can be seen only through microscopes are called micro-organisms. However, the most practical method is to classify soil organism into plants or animals to which they belong. Accordingly, soil organisms are grouped into the follow- I: ing two categories, namely (A) soil fauna and (B) soil i, flora.

Some of the important soil organisms which play an irreplaceable role in recycling of plant nutrient, growth and development of plants are discussed here.

Soil Fauna

This group includes organisms belonging to animal kingdom, e.g. earthworm, termite, nematode, springtail, mite, protozoa, enchyatraeid worms, mesofauna, larvae of beetles and dipterous flies, ants, amoeba, gastropods : and soil inhabiting mammals.

Earthworm

Earthworms form the most important macrofauna of the soil. More than 200 species of earthworms have been identified the world over, but in India *Pheretima posthuma* is the most commonly found earthworm species. In some places other species are dominating. The length of the worms varies from a few centimeters to even 3 metre depending on the species.

In the upper 15-25 cm of soil, earthworms play an important role of working the soil. They eat soil and the amount of soil these creatures pass through their bodies annually may amount to as much as 37.5 tonnes/ha of dry earth, a startling figure. During the passage through the alimentary canal of the worms, not only the organic matter that serves as food but also the mineral constituents are subjected to digestive enzymes and to a grinding action. Earthworm cast on the cultivated field may weigh as much as 18000 kg/ha. Compared to the soil itself, the casts are definitely higher in bacterial contents, organic matter and other nutrient. Luxuriant growth of grass around earthworm casts suggests an increased availability of plant nutrients therein. Earthworms are regarded as farmers friends due to their noticeable favourable effect on soil productivity.

Besides leaving fertile casts in the soil, earthworms burrow holes in the soil which improve aeration and drainage. They mix and granulate the soil by dragging into their burrows quantities of un-decomposed organic matter such as leaves and grasses which they use as food. In some cases, the accumulation is surprisingly large. II In uncultivated soils this is more important than in ploughed land, where organic matter is normally turned over 1 in large quantity. Without any doubt, earthworms in- I: crease both the size and stability of the soil aggregates.

A well aerated and moist habitat is ideal for the I growth of earthworms in soil. They are found mostly in medium-textured upland soils where the moisture capacity is high rather than in draughty sands or poorly drained lowlands. They must have organic matter as a source of food. Temperature also affects earthworm number, a temperature of 10-150 C is most suitable. In very cold weather they penetrate into the soil as deep as one to two metres, to find a suitable temperature zone. In barren soil, a sudden heavy frost may kill earthworms in large number on the soil surface.

Because of their sensitivity to soil and other environmental factors, there is a wide variation in the number of earthworms in different soils. In very acid soils under conifers, an average of fewer than one organism per square metre is common. In contrast, more than 500 per square metre have been found on rich well-granulated soils. The number commonly found in arable soils range from 30 to 300 per square metre equivalent to 3 lakh to 3 million per hectare furrow slice. The biomass or live weight for this number would range perhaps 110-1100 kg/ha.

Termites

Termites are white coloured small and most commonly occurring insect in the soil. They play an important role in breaking down of organic material occurring in the soil. Tropical and subtropical climate is very congenial for their growth. Thus they are profusely found in Indian soils. In tropical grasslands of savannas and tropical forest lands they are very prominent. They carryon a very complex social life in their honeycomb like houses or mounds, which serve essentially as their 'cities'. For constructing their houses (habitat) they transport soil from lower layers to above the surface soil level thereby bringing about extensive mixing of soil materials and of the plant residues they use as food. The quantities of materials they deposit often compare favourably with the surface castings of earthworms amounting to tens or even hundreds of tonnes per hectare at any given time. The heights of their mounds sometimes go as high as 6 metres or more and may extend to an even greater depths into the soil in some desert areas. Each mound may house millions of termites.

Though the overall contribution of termites to soil productivity is generally less than that of earthworms but their digestive system is generally more efficient. Micro-organisms such as bacteria and protozoa in the gut of the termites readily attack cellulose and other complex organic materials, releasing sugars and other nutrients for use by termites. This symbiotic process is helpful : in rapidly breaking down woody materials.

Termite deposits commonly have a lower organic matter content than the surrounding undistributed top soil. This may be due to the fact that these deposits contain much lower soil organic subsoil material, which was brought to the surface as the mounds were being formed. Crop growth in soils from areas where these mounds have existed is often poor not only because of low nutrient content in the surface layers, but because of the greater compactness of the mound material, the particles of which had been cemented together by the termites as the mound was constructed.

Besides their useful roles like breakdown of organic matter, termites may also cause negative effects on plants. They can eat seeds and roots of the plant, thereby causing damage to plant. However, their positive roles far out-weigh the negative roles.

Nematodes

Nematodes are also called thread worms or eelworms, and they are found in soil in large numbers. These organisms are round or spindle shaped, the ends of the worm's body being pointed. They are very ,minute in size and can i be seen only with the help of a microscope. The food t habit of the nematodes is variable, some live on decaying : organic matter; some are predatory on other nematodes, bacteria, algae, protozoa and the rest are parasitic, attacking the roots of higher plants to pass at least a part of their life cycle embedded ill such tissue. Saprophytes (feeding on decaying organic residues) and predators (eating other soil organisms including nematodes themselves) are by far the most numerous in the average soil and most varied. The importance of nematodes lies in their relationship with other soil organisms.

Nematodes are both useful and harmful. Useful in the sense that they help in decomposition of organic residues of the soil and control the growth of other soil organisms which they may otherwise prove damaging to crop plants. They are harmful when they are parasites on crop plants and create serious problem.

Protozoa

Protozoa form a group of very primitive, single celled organisms ranging in size from several microns (1 mm = 1000 microns) to one or more centimeters. They are al- most universally found in arable lands. In highly adverse conditions like lack of moisture or food, and too low or too high temperature conditions etc. they form a protective shield around their body called cyst, which help them to survive the adverse weather. On finding favourable atmosphere, cysts from around the cell body dissolves and the organism again becomes active.

A few protozoa can synthesize their own food as they contain some pigments to utilize solar energy. However, majority of them are saprophytic or holozoic being characterized by direct feeding upon microbial cells. They feed largely upon bacteria and other protozoan strains. In holozoic nutrition, cell of organisms, dead or alive, are as such surrounded by protozoa and ultimately ingested into the body. These cells serving as food are mostly bacteria. It is estimated that from formation till beginning of reproduction, protozoa may consume as much as 40 thousand cells of bacteria. This shows that protozoa keep a check on unduly high growth of bacteria by consuming the excess bacterial cells, thereby maintaining a bio- logical equilibrium in the cell.

Protozoa population in the soil fluctuate rapidly, aeration and food supply being the most important factors influencing It. Most of the organisms thrive best under conditions of good aeration and therefore, confined to the surface horizons. Usually their numbers are highest in the spring and autumn seasons. Population of a few thousands to several hundred thousand per gram of the upper few centimeters of soil have been observed.

The common figure is 10,000 to 100,000 per gram. This might amount to a live weight of as much as 170 or even 200 kg to the hectare furrow slice.

Rotifers

About 100 species of rotifers have been identified so far. They grow well in moist condition especially in swampy lands where their numbers may be surprisingly high. These organisms are mostly microscopic in size. The anterior is modified into a retractile disk bearing cilia (minute hair) which in motion, give the appearance of moving wheels. These cilia or hairs attract floating food materials and cells absorb them for nutrition. The importance of rotifier to crop plants is not yet fully under- stood. However, it is known that their activities are largely confined to peat, marshy and in wet places occurring in mineral soils.

Encbytraeid Worms

These worms have not received much attention probably because they are found in soils of low agricultural value. They are mostly found in grasslands and coniferous forest soils. These soils contain between 10,000 to 300,000 worms per square metre area, with a live weight of 30 to 100 kg/ha. They consume rotting material and may feed on the bacteria and fungi.

Mesofauna

This group largely comprises of mites that vary in size from 0.1 to 1 mm. They are present in soil in considerably large numbers belonging to a large number of species. Each species probably has specific food requirements but the mites as a whole feed on a wide range of materials such as decomposing plant remains and fungi though perhaps many of them appearing to feed on plant remains, are in reality feeding upon the micro-organisms carrying out decomposition and some are predacious on i other members of mesofauna. The springtail and collembola are rather larger, 0.5 to 2 mm in size, wingless insects. Their number in soil in comparison to mites is very less. Their feed on decaying plant tissues and micro-organisms decomposing it, on dead insects and presumably on remains of other members of the soil fauna. The species present in the soil vary with the depth at which they live, the soil surface inhabitants being pigmented, and have well developed and springing organs, all of which are lacking or poorly developed in the subsoil dwellers. The mites and springtails are extremely found in forest soils. The number of mites and springtails is lower in arable soils than in undisturbed soils. Ploughing the grasslands reduces their number. The actual number in an arable soil is dependent upon the organic matter contents of the soil. In the soils regularly receiving compost or organic manure, number is higher than in those receiving only chemical fertilizers.

Larvae of Beetles and Dipterous Flies

Larvae is one of the stages in the life cycle of beetle. Larvae of many beetles and other flies live in the soil and can be important borrowers in the soil. They can also attack roots of crop plants or the young plants just above the ground level. Among the principal agricultural pests on living crops are wire-worms which are the larvae of elaterid beetles.

Ants

Ants are also very efficient insects in working the soil and fostering granulation and aeration. They collect green leaf strains, and even seeds of grass as their food. Functionally they are similar to termites but they differ from termites in that they are not confined to equatorial and subtropical regions but are common inhabitants of temperate soils also. They can also differ in their feeding habits. Many genera of ants are predacious (eat other ii kind of ants) and feed on the mesofauna and termites. Ants are not confined only to soil fauna but they can also feed on insect pests.

Myriapods and Isopods

They include millipedes and centipedes which are found in soil in considerable number. In general the food of the millipedes is decaying plant litter and centipedes are predacious on the immature form of the larger soil fauna. Some millipedes are unable to digest cellulose, so must consume a very large amount of litter to extract sufficient sugars and other simple carbohydrates for their requirements, hence they can be very important as mechanical comminuters and mixers of litter with soil.

Atmospheric humidity considerably influences the number of these two insects. Many species have little ability to control water loss from their body atmosphere and many cannot control water intake if placed in water. A good forest mull or the surface layers of the forest soils with leaf litter forms a very favourable environment for most species. Hence forest soils that are not too acid but are, for some reasons, not suitable for higher earth- worm populations, such as sandy soils, usually have a high millipede population.

In agriculture, millipedes are found in some pastures and many occur in considerable numbers in arable soil well supplied with organic manures and these may be so high as to become pest of some crops such as potatoes.

Gasteropods

Slugs and snails are the two soil representatives of the gasteropods. Very little is known about their distribution and number. Most are surface feeders, usually active in damp conditions, hence during the day either burrow into the soil or into dense shady spots such as under stones or leaves or in thick grass tussocks. Some species of snails can digest cellulose easily. Thus the forest floor is a well-adapted environment for them and litter and fungi appear to be their principle source of food.

Slugs usually feed on dying vegetation such as freshly fallen leaves, old grass and so on and that they usually attack the actively growing parts of plants only when other foods are scarce. They are, thus typical scavengers rendering old plant tissues more usable by the smaller members of the animal population by excreting them in a marcerated and partially digested form. Some slugs are predacious on other slugs, worms and centipedes. Normally, the number of slugs on arable land is not very high, though in some areas they may damage potatoes. Their number in a wheat field is found to be 15 laths per hectare weighing about 400 kg.

Soil Flora

This is the group of organisms which belong to plant I' kingdom. Prominent members of this group are bacteria, algae, fungi, actinomycetes and virus.

Bacteria

Bacteria are a large group of single celled microscopic organisms, widely distributed in soil, air, water, animal body, plant tissues and even foods. In the soil bacteria form the largest microbial population and are of great, ecological, biological and agricultural significance.

Distribution and Abundance

The various methods of counting bacterial population ! give variable estimates. However, the number of bacteria present in soil are also variable as many conditions affect their growth. In general, greatest population is in the surface soil since conditions of temperature, moisture, aeration and food are more favourable. The number of .bacteria ranges from a few billion to 3 trillion in each kilogram of soil. A biomass of few hundred kilogram to 5 tonnes live weight to the hectare furrow slice of fertile soils is commonly encountered. In each gram of a representative soil, bacterial number may range from several i hundred thousands to one hundred million depending on \ the environmental conditions.

The number and type of bacteria are governed to a large extent by soil type and cultivation practices. The population in most grasslands, for example, is greater than in arable land, because of the greater root density and the larger supply of the utilizable organic matter coming from root decomposition and plant debris. Again bacterial numbers are higher in cultivated than in virgin land.

Cultivation usually makes conditions more favourable for bacterial proliferation.

Environmental Influences

The density and composition of bacteria in the soil is greatly affected by environmental conditions. The primary environmental factors influencing soil bacteria include moisture, aeration, temperature, organic matter, acidity and inorganic nutrient supply. Many other subsidiary factors such as cultivation, season and depth also affect bacterial population of the soil as briefly discussed earlier.

Moisture is one of the important factors governing microbial activity. Since water is the major component of cells (protoplasm), an adequate supply must be available for vegetative development. But where moisture becomes excessive, microbial proliferation is suppressed because of limited supply of oxygen and the environment around the bacteria become deficient in oxygen. The maximum bacterial density is found in regions of fairly high moisture content, and the optimum level for the acti- vities of aerobic bacteria is at 50-75 % of the soil moisture holding capacity.

Temperature governs all biological processes and thus it also strongly affects the trend of fluctuation in bacterial population of the soil. An association between population size and temperature is well studied. Each micro-organism has an optimum temperature range for growth, outside of which development ceases. Based on the temperature requirements, bacteria are grouped into three categories namely psychrophilic, mesophyllic and thermophyllic. These bacteria which grow best at temperature below 200 C are called psychrophilic. Most micro-organisms are mesophyllic for which the optimum temperature for proliferation ranges from 20-450 C. The bacteria flourishing the best under temperature range of 45-65°C are termed thermophyllic. In the soil, hardly any bacteria is found in its active form beyond 650 C temperature as the soil itself does not reach such a high level of temperature. However, a fourth term is also used, for laboratory studies for the bacteria growing above 650 C temper rature, namely thermoduric.

Organic matter is yet another important factor influencing soil bacterial population. Localities in the field I having rich humus content have the largest bacterial number.

Highly acidic or alkaline conditions tend to inhibit many common bacteria as the optimum for most species is near neutrality. The greater the acidity the smaller generally is the size of the bacterial population. The " liming of acid soils, generally gives greater bacterial counts.

Nutrition

Bacteria is a highly versatile organism. One or the other species of bacteria can decompose more or less all the substances found in the soil.

Barring a very few species most of the bacteria derive I their food from organic matter. Carbon and energy, are the two basic and immediate requirements of bacteria. Some bacteria derive their carbon from atmospheric carbon dioxide and energy from sunlight and some others derive carbon by degrading organic matter of the soil and energy from oxidation of organic matter. The latter category forms the majority. To acquire all the essential substances needed to synthesize the body of bacteria, a huge amount of organic matter is consumed. Though many other micro-organisms playa key role in decomposition but bacteria occupy the most prominent position. This conversion is necessary to maintain an ecological equilibrium in the nature.

Bacteria can attack all the principal components of an organic material namely protein, fat, carbohydrates, minerals and vitamins to meet their food and energy requirements. Besides, they carry out numerous biochemical reactions in the soil which are essential.

Fungi

Fungi is a primitive group of organisms, which play a very important role in transformation of soil constituents. In terms of number they are much less than bacteria in the soil. However, in most well aerated cultivated soils, the fungi account for the largest part of the total microbial content. The body of fungi is grown mostly in an extensive network of filaments called hyphae. In forest soils where organic matter accumulates in surface horizons, fungi are prominent and acid conditions in general are favourable for fungal growth.

The filaments of fungal body are not visible to the naked eyes but they may be seen by direct examination of soil with microscopes. Fertile land may contain as much as 10 to 100 million of active fungal filaments per gram of soil.

Distribution and Abundance

To estimate the fungal population, is an extremely difficult task because even a minute fragment of the body filament of fungi may give rise to a new clamp of fungi.

However, various methods deployed so far to reveal their i presence in soil is population typically ranging from as few as 20,000 to as many as 1,000,000 fungal units per gram of soil, the unit being considered as spore, hyphae, hyphal fragment etc. which are capable of giving rise to a colony in artificial medium. Colony is the term used Ii for a dense growth as marked by spots in the petridish containing nutrient medium. This may amount to a ii fungal biomass of about 500 kg in a hectare of area.

For convenience of study, fungi are divided into three groups: (i) yeasts, (ii) moulds, and (iii) mushroom fungi. The distinctly filamentous and microscopic moulds have a role infinitely more important in the soil than the mush- room fungi, approaching or even exceeding sometimes the influences of bacteria. The greater number of moulds are found in the surface layers where organic matter is ample and aeration adequate. Their numbers fluctuate greatly with soil conditions; perhaps 1 lath to 1 million individuals in one gram of dry soil. This may amount to a range in biomass of 800-8000 kg/ha. The population is highly fluctuating.

Environmental Influences

The population of fungi at a time is determined by soil and climatic conditions. Both the generic composition and the size of the flora vary with the type of soil and with its physical and chemical characteristics. Whether a given micro-organisms will be able to survive, adapt itself and become established in the given habitat will be determined by the surrounding environment. The major external influences imposed upon the fungus flora include the organic matter status, soil reaction, fertilizers and manure, soil moisture, aeration, temperature, depth, season of the year and the composition of the vegetation.

None of the fungi contain any kind of pigment in their body, thus they cannot synthesize their own food with the help of sunlight or atmospheric carbon dioxide unlike algae and a few other bacterial species. Therefore, fungi solely depend upon organic matter of the soil for their nutrition, and the spots where organic matter is accumulating in higher amount, fungal population is also higher. Organic matter content of the soil is regarded as the major determinant of fungal population and a positive correlation is found between the humus content of soil and size of fungal population. Addition of farmyard manure, compost, green manure and other bulky organic manures result in marked increase in fungal population of the soil.

Second most ecological factor affecting fungal growth is soil reaction. Many mould species can flourish over a wide pH range from a highly acidic to the alkaline extremes. However the acid conditions (pH below 7) seem to favour the growth of fungi. Because bacteria and other organisms are uncommon in acid soils, the microbial population in such soils is dominated by fungi. This is regarded not as the result of the fungi finding their optimum in acidic conditions but rather it is a consequence of the lack of microbiological competition for the food reserve. Thus the insensitivity of many of the moulds to acidity and the narrow pH range of most other micro- organisms lead to the conclusion that the fungi make up a large percentage of the population and are responsible for a considerable portion of the biochemical transformations in acidic soils. The other ecological factors affect the growth of fungi in the similar way as in case of bacteria.

Nutrition

As already discussed the sole source of food for fungus is organic matter. Fungi derive carbon and energy from easily oxidizable organic matter. This kind of nutritional, habit provides a great importance to fungi for carbohydrate decomposition. Cellulose, starch, gums, lignin and sometimes proteins and sugars succumb to their attack.

Moreover, fungi function more efficiently than bacteria in that they transform into their tissues a large pro- I portion of decaying plant residues and give off as byproducts less carbon dioxide and ammonium. Upto half of the substance decomposed by moulds may become organism tissues, compared to about 20 % for bacteria.

Though fungi can degrade the protein to the extent bacteria can, yet the soil fertility depends in no small degree on moulds, since they continue the decomposition process after bacteria and actinomycetes have essentially ceased to function.

Certain types of fungi associate themselves with the roots of higher plant and this association is called mycorrhiza. This is symbiotic association in which both the fungi and the roots benefit each other. Extensive fungal mycellium gets a stable shelter with plant roots by lacing around or penetrating the plant roots and the mycelial portion outside the plant roots which remain embedded in soil, absorbs nutrients, especially phosphorus from the soil and carries it inside the plant roots. Plant is thereby nourished with phosphorus.

Algae

The algae is also, like fungi, a group of primitive or under developed plants. However, they contain a pigment called chlorophyll which enables them to synthesize their own food in the presence of sunlight using atmospheric carbon dioxide. They perform best under moist to wet conditions. However, a few forms of algae obtain their energy largely from organic matter and live within and below the surface of the soil. Several hundred species of algae have been isolated from soils. Soil algae are divided into four general groups, namely, (a) blue green algae, (b) green, (c) diatoms and (d) yellow-green.

The green algae are most evident in soil, especially if the pH is low. In vegetative form, soil algae are prominently present on the surface soil. In sub-soils most algae are present as resting spores or cysts or in vegetative forms that do not depend upon chlorophyll. Grassland seems especially favourable for the blue-green forms whereas in old gardens diatoms are most numerous. All the ordinary types of algae are greatly stimulated by the application of farm manure.

Both, the green and blue-green algae out number the diatoms. A common range in algal population is from 1 to 10 billion per square metre at 15 cm deep (10,000- 100,000 per gram). Blue-green algae are especially numerous in rice soils, and when such lands are flooded, appreciable amounts of atmospheric nitrogen are fixed or changed to a combined form by these organisms. Blue-green algae growing within the leaves of the aquatic form azolla are also able to fix nitrogen in quantities significant for rice production.

Algae can be found growing around the mouths of wells. They are abundant in habitats in which moisture 1 is adequate and light is accessible. Their development on the cultivated and virgin land is frequently noted with the naked eye but other minute forms can be obtained : from lower depths as well. Their presence can be demonstrated by the addition of small amount of soil to a mixture of chemical substances containing potassium, magnesium, phosphate, sulphate, nitrate, calcium and iron and i traces of other inorganic nutrients. The resultant growth is visible to unaided eye as green colour mass. The distinctive green pigmentation associated with these minute plants result from their possession of chlorophyll.

Algae not only supplies nitrogen to the soil but also supplies a pretty good amount of organic matter too. By virtue of being bestowed with a natural capacity of synthesising their own food, their body growth is much bulkier than that of other micro-organisms. The cycle of their reproduction, death and decay continues in nature thereby enriching soil in both nitrogen and organic matter.

Actinomycetes

There is a transition group of micro-organisms between the simple bacteria and the fungi. The group actinomycetes overlaps its more primitive (bacteria) and its more developed (fungi) neighbour. Numerically their presence in the soil is next to bacteria. Actinomycetal body has certain morphological characters which are common in both: bacteria and fungi. Thus, we can say that the actinomycetes is a in-between stage of bacterial and fungi in terms of evolution.

Actinomycetes develop best in moist, well-aerated I and slightly alkaline soil. But in times of drought, they 1 remain active to a degree not generally exhibited by either l bacteria or moulds. They are rather sensitive to acid conditions, their growth being practically prohibited in mineral soils at a pH of 5.0 or below. Their optimum development occurs at pH values between 6.0 and 7.5. This relationship between pH and growth of actinomycetes is used for controlling scab disease of potato. In alkaline soil, potato may suffer from scab disease and the disease can be controlled by using sulphur to the field which brings down the soil pH.

Population density of actinomycetes in the soil is [next to bacteria, their number sometimes reaching hundreds of millions, one tenth the number of the bacteria. In actual live weight they often exceed bacteria. Under, especially favourable conditions, a biomass of more than 4500 kilograms of actinomycetes threads and spores might be present in a hectare furrows slice. These organisms are especially numerous in soil rich in humus such as meadows or pastures where the activity is not too great. There they sometimes exceed in numbers to all other micro-organisms. The addition of farmyard manure or compost markedly stimulates their growth. The aroma of freshly ploughed or freshly irrigated land that is so noticeable at certain times of the year is probably due to actinomycetes as well as to certain moulds.

Actinomycetes are also of great importance in the decomposition of organic matter and the liberation of nutrients. Apparently they reduce to simpler forms even the more resistant compounds such as cellulose, chitin and phosphatic compounds. The presence of actionomycetes in abundance in soils long under sod is an indication of their capacity to attack complex com pounds.

General Ecology of Soil Organisms

Interactions and interrelations of the various life) forms are the hallmark of soil chemistry. However, \ these are always in a continued stage of change. The absence or dominance of an organism is a result of associations and interactions of all individuals found in the It population. Environmental changes temporarily upset the equilibrium, but it is re-established, possibly in a modified form, as the population tend to shift in accordance with new circumstances. The ecological inter relationship among the microorganisms have various manifestations which are discussed under the following heads.

Competition Among Soil Micro-organisms

An intense rivalry exists among soil micro-organisms for food and air thereby submitting to the principle of survival of fittest wherein some micro-organisms succeed to establish themselves while the others are suppressed. When fresh organic matter is added in the soil, the vigorous strains of bacteria, fungi and actinomycetes compete with each other for this source of food. The bacteria dominate initially since they can reproduce rapidly and prefer simple compounds. As these simple compounds are broken down the fungi and particularly the actinomycetes become more competitive. Undoubtedly such competition for food is the rule and not the exception in soils.

Besides the food competition, there is another kind i of microbial rivalry just as intense and even more deadly. Certain bacteria, fungi and actinomycetes can produce , chemical substances that will inhibit or even actually kill other microbes. These chemical substances are called antibiotics and their use is now extensively made for controlling human, animal and plant diseases.

The organisms which feed on other organisms are called predators and parasites. Protozoa are the commonest predators on bacteria. Protozoa considerably reduce the bacterial population in the soil by consuming millions of bacteria. This is necessary to maintain a biotic equilibrium otherwise the rapidly growing bacterial population may pose a serious threat to the existence of all other micro-organisms. However, on the other hand protozoa themselves are subjected to predation by other organisms. Thus the bacterial populaton is never sup- pressed by protozoa beyond a reasonable limit.

Certain bacteria and actinomycetes can parasitise fungi. On the other hand, certain genera of fungi can feed upon other genera of fungi themselves. Thus, the nature has established very complex but useful devices which keep a check on excessive growth of all the kinds of organisms, thereby maintaining a biotic equilibrium.

Harmful Effects of Organisms on Plants

These effects are the results of interaction between plants and organisms. Certain macrofauna viz. rodents, moles and termites may greatly damage crops. Snails and slugs in some climates are dreaded pests especially of vegetables. Ants, as they transfer aphids on certain plants must be diligently combated by farmers. Also most plant roots are infested with nematodes sometimes so seriously the successful growth of certain crops becomes difficult and expensive. Tobacco, potato and cotton are examples of the crops adversely affected by nematodes. Crop rotation and development of resistant varieties are the most effective means of controlling these pests.

Most of the plant diseases are caused by microflora like bacteria, fungi and actinomycetes. They all exert I; a highly devastating impact on higher plants. Fungi are responsible for most of the soil-borne diseases of crop plants. Some of the more common diseases produced by soil flora are wilts, damping off, root rots, club root of cabbage and similar crop and actinomycetes scab of potato. Injurious organisms live for variable periods in the soil. Some of them disappear in a few years time if the plants upon whom they grow (host plants) are not grown in such soil, but others are able to live for years so long as they find a little organic matter in the soil. Infection is easy and organism may disseminate through agricultural instruments, people, cattle, plants etc.

These organisms, snatch away the food of plants, cause decay of plant parts when the plant is still growing and destabilize the whole physiology of the plant. This injurious interaction assumes the form of plant diseases.

Beneficial Effects of Soil Organisms

Soil organisms play an irreplaceable role in the growth and development of crop plants. Certain important roles are discussed here.

Organic Matter Decomposition

The food chain of the living things consists of the green-chlorophyll bearing plants which manufacture their own food material (photosynthesis). These plants or their residues form the food supply for the herbivores (plant eating animals) which in turn become the prey for the carnivores (flesh eating animals).

Ultimately, all plant and animal residues are returned to the soil and subjected to decomposition by micro- organisms. As a result of these microbiological processes, the elements which were originally consumed by plants are returned to circulation for re-utilization. Fortunately, plant and animal residues do not undergo rapid and complete mineralization. A large proportion of these organic deposits are first modified by these organisms to form resistant brown or black complexes, which in combination with constituents of microbial cells and other resistant products of metabolism form the humus. Thus humus consists of unavailable form of carbon, nitrogen, phosphorus, sulphur and potassium, and acts as a reservoir. Its gradual decomposition of micro-organisms results in the liberation of a slow but continuous stream of elements essential for new plant synthesis. In this way the major cycle of life is completed, from the soil back to .the soil.

Inorganic Transformation

Micro-organisms are gifted with a natural capacity to convert organic compounds into inorganic compounds and inorganic ones to organic. Organic compounds which are found in the plants or animals or their remnants are not available to the plant roots. Micro-organisms of the soil convert the organic compounds of organic matter into inorganic forms thereby making them avail- able to plant roots.

However, if needed micro-organism can again convert inorganic forms of compounds into organic forms by directly consuming it. For example, when an organic residue is added in the soil its protein and organic compound are broken down by micro-organisms to ammonium and nitrate-both inorganic compounds. If crop is growing it can absorb these two. On the other hand, formation of other bacterial cells itself requires all these easily available inorganic compounds. Thereby they keep the nutrients preserved in the soil.

Nitrogen Fixation

Nitrogen is the most important plant nutrient and is required by the plants in heavy doses. Though, the atmosphere contains -78 % nitrogen by volume, yet plants cannot utilize it unless it is converted into solid compound forms. Conversion of nitrogen from gas to solid com- pound form is called nitrogen fixation. Certain micro- organisms living in the soil and plant roots have capacity to absorb atmospheric nitrogen directly. This nitrogen is ultimately availed to the crop plants after undergoing a series of biochemical changes. This capacity of micro- organisms is extremely important for nitrogen nutrition of plants.

Some algae, (e.g. blue-green) bacteria (e.g. Rhizobium, Azotobactor) and actinomycetes are significant nitrogen fixing organisms.

The discussion made above shows that the organisms of the soil must have energy and nutrients if they are to function efficiently. In obtaining them they break down organic matter aid in the production of humus and leave behind compounds that are useful to higher plants.

3

Mineral Matter in Plants

Minerals are required by plants as part of their food, to form their structure. The firmness of straw for example, is due to the presence in it of silica, the principal constituent of sand and flints. Potassa, soda, lime, magnesia, and phosphoric acid are contained in plants, in different proportions. All of these they must obtain from the soil. The alkalies above named appear to be essential to the proper development of the higher vegetable forms. Some plants require them in one mode of combination, and some in another; and thus the soil that is very good for one, may be quite unfit for others. Firs and pines find enough to support them in barren, sandy soil.

The proportion of silicate of potash, (necessary for the firmness of wheat straw), does not vary perceptibly in the soil of grain fields, because what is removed by the reaper, is again replaced by decaying straw. But this is not the case with meadow-land. Hence you would never find a luxuriant crop of grass on sandy and limestone soils which contain little potash, evidently because one of the constituents indispensable to the growth of the plants is wanting. If a meadow be well manured, we remove, with the increased crop of grass, a greater quantity of potash than can, by a repetition of the same manure, be restored to it. So grass-land manured with gypsum soon ceases to feel its agency. But if the meadow be strewed from time to time with wood ashes, or soap-boilers' lye made from wood ashes, then the grass thrives as luxuriantly as before. The ashes are only a means of restoring the necessary potash for the grass stalks. So oats, barley, and rye may be made for once to grow upon a sandy heath, by mixing with the scanty soil the ashes of the heath-plants that grow upon it. Those ashes contain soda and potash, conveyed to the growing furze or gorse by rain-water. The soil of one district consists of sandstone; certain trees find in it a quantity of alkaline earths sufficient for their own sustenance. When felled, and burnt and sprinkled upon the soil, oats will grow and thrive that without such aid would not vegetate.

The most decisive proof of the absurdity of the indiscriminate use of any strong manure was obtained at Bingen, a town on the Rhine, where the produce and development of vines were highly increased by manuring them with animal matters such as shavings of horn. After some years, the formation of the wood and leaves decreased perceptibly. Such manure had too much hastened the growth of the vines: in two or three years they had exhausted the potash in the formation of their fruit leaves and wood; so that none remained for the future crops, as shavings of horn contain no potash. Cow-dung would have been better, and is known to be better.

Nutrient Pollution

Nutrient pollution, a form of water pollution, refers to contamination by excessive inputs of nutrients. It is a primary cause of eutrophication of surface waters, in which excess nutrients, usually nitrogen or phosphorus, stimulate algal growth. Sources of nutrient pollution include runoff from farm fields and pastures, discharges from septic tanks and feedlots, and emissions from combustion.

Excess nutrients, or nutrient pollution, have been summarized as potentially leading to:

- Population Effects: excess growth of algae (blooms);
- Community Effects: species composition shifts (dominant taxa);
- Ecological Effects: food web changes, light limitation;
- Biogeochemical Effects: excess organic carbon (eutrophication); dissolved oxygen deficits (hypoxia); toxin production;
- Human health effects: excess nitrate in drinking water (blue baby syndrome); disinfection by-products in drinking water

In a 2011 EPA report, the Science Advisory Board succinctly states: "Excess reactive nitrogen compounds in the environment are associated with many large-scale environmental concerns, including eutrophication of surface waters, toxic algae blooms, hypoxia, acid rain, nitrogen saturation in forests, and global warming."

Excess Nutrients and TMDLs

The regulatory mechanism in the United States, a Total Maximum Daily Load (TMDL), prescribes the maximum amount of a pollutant (including nutrients) that a body of water can receive while still meeting U.S. Clean Water Act water quality standards. Specifically, Section 303 of the Clean Water Act requires each state to generate a TMDL report for each body of water impaired by pollutants. TMDL reports identify pollutant levels and strategies to accomplish pollutant reduction goals. EPA has described TMDLs as establishing a pollutant budget then allocating portions of the overall budget to the pollutant's sources. For many coastal water bodies, the main pollutant issue is excess nutrients, also termed *nutrient over-enrichment*. A TMDL can prescribe the minimum level of Dissolved Oxygen (DO) available in a body of water, which is directly related to nutrient levels. In 2010, 18 percent of TMDLs nationwide were related to nutrient levels including organic enrichment/oxygen depletion, noxious plants, algal growth, and ammonia. In Long Island Sound the TMDL development process enabled the Connecticut Department of Environmental Protection (CTDEP) and the New York State Department of Environmental Conservation (NYSDEC) to incorporate a 58.5 percent nitrogen reduction target into a regulatory and legal framework.

Nutrient Remediation

Innovative solutions have been conceived to deal with ecological disturbances caused by nutrient pollution, especially in aquatic systems. Solutions generally alter or enhance natural processes to shift nutrient effects away from detrimental ecological impacts, usually by eliminated reactive nutrients. Nutrient remediation is a form of environmental remediation, but concerns only biologically active nutrients such as Nitrogen and Phosphorus. "Remediation" refers to the removal of pollution or contaminants, generally for the protection of human health. Nutrient removal technologies include biofiltration, which uses living material to capture and biologically degrade pollutants. Examples of biofiltration include: green belts, riparian areas, natural and constructed wetlands, and treatment ponds. These areas most commonly capture anthropogenic discharges such as wastewater, stormwater runoff, or sewage treatment, for land reclamation after mining, refineries, or land development. Biofiltration utilizes biological assimilation to capture, absorb, and eventually incorporate the pollutants (including nutrients) into living tissue. Another form of nutrient removal is bioremediation, which uses micro-organisms to remove pollutants. Bioremediation can occur on its own (natural attenuation or intrinsic bioremediation) or can be encouraged by the addition of fertilizers (biostimulation).

"Nutrient bioextraction" is the preferred term for bioremediation involving cultured plants and animals. Nutrient bioextraction (also called bioharvesting) is the practice of farming and harvesting shellfish and seaweed for the purpose of removing nitrogen and other nutrients from natural water bodies. It has been suggested that nitrogen removal by oyster reefs could generate net benefits for sources facing nitrogen emission restrictions, similar to other nutrient trading scenarios. Specifically, if oysters maintain nitrogen levels in estuaries below thresholds that would lead to the imposition of emission limits, oysters effectively save the sources the compliance costs they otherwise would incur. Several studies have shown that oysters and mussels have the capacity to dramatically impact nitrogen levels in estuaries.

History of Nutrient Policy in the United States

Policy for a National Nutrient Strategy was created 1998, focused on developing nutrient criteria. Criteria, for Rivers/Streams, Lakes/ Reservoirs, Estuaries, Wetlands, and guidance were completed between 2000-2010, including "Ecoregional" Nutrient Criteria in 14 ecoregions across the U.S. In 2004 the EPA Office of Science and Technology defined EPA's expectations for numeric criteria for total Nitrogen (TN), total Phosphorus (TP), Chlorophyll a(chl-a), and clarity, and established "mutually-agreed upon plans" for state criteria development. In 2007 EPA reiterated EPA's expecta tions for numeric criteria and committed EPA to support State efforts.

Trading and Offsets

Interest in nutrient removal, and achieving regional TMDLs, has led to the development of nutrient trading schemes. Nutrient trading is a type of water quality trading, which is a market-based policy instrument used to improve or maintain water quality. Trading is based on the fact that sources in a watershed can face very different costs to control the same pollutant. Water quality trading involves the voluntary exchange of pollution reduction credits from sources with low costs of pollution control to those with high costs of pollution control, and the same principals apply to nutrient water quality trading. The "polluter pays" principle is often an underlying force, usually with a linked regulatory driver for participation is the trading programme. Forest Trends 2013 report summarized water quality trading programmes and found three main types of funders: beneficiaries of watershed protection, polluters compensating for their impacts and 'public good payers' that may not directly benefit, but fund the pollution reduction credits on behalf of a government or

NGO. As of 2013, payments were overwhelmingly initiated by public good payers like governments and NGOs.

Soil pH

The soil pH is a measure of the acidity or basicity in soils. pH is defined as the negative logarithm (base 10) of the activity of hydronium ions (H+ or, more precisely, H_3O^+aq) in a solution. It ranges from 0 to 14, with 7 being neutral. A pH below 7 is acidic and above 7 is basic. Soil pH is considered a master variable in soils as it controls many chemical processes that take place. It specifically affects plant nutrient availability by controlling the chemical forms of the nutrient. The optimum pH range for most plants is between 5.5 and 7.0, however many plants have adapted to thrive at pH values outside this range.

Classification of Soil pH Ranges

The United States Department of Agriculture Natural Resources Conservation Service, formerly Soil Conservation Service classifies soil pH ranges as follows:

Denomination	*pH range*
Ultra acid	< 3.5
Extreme acid	3.5–4.4
Very strong acid	4.5–5.0
Strong acid	5.1–5.5
Moderate acid	5.6–6.0
Slight acid	6.1–6.5
Neutral	6.6–7.3
Slightly alkaline	7.4–7.8
Moderately alkaline	7.9–8.4
Strongly alkaline	8.5–9.0
Very strongly alkaline	> 9.0

Sources of soil pH

Sources of Acidity: Acidity in soils comes from H^+ and Al^{3+} ions in the soil solution and sorbed to soil surfaces. While pH is the measure of H^+ in solution, Al^{3+} is important in acid soils because between pH 4 and 6, Al^{3+} reacts with water (H_2O) forming $AlOH^{2+}$, and $Al(OH)_2^+$, releasing extra H^+ ions. Every Al^{3+} ion can create 3 H^+ ions. Many other processes contribute to the formation of acid soils including rainfall, fertilizer use, plant root activity and the weathering of primary and secondary soil minerals. Acid soils can also be caused by pollutants such as acid rain and mine spoilings.

- Rainfall: Acid soils are most often found in areas of high rainfall. Excess rainfall leaches base cation from the soil, increasing the percentage of Al^{3+} and H^+ relative to other cations. Additionally, rainwater has a slightly acidic pH of 5.7 due to a reaction with CO_2 in the atmosphere that forms carbonic acid.
- Fertilizer use: Ammonium (NH_4^+) fertilizers react in the soil in a process called nitrification to form nitrate (NO_3^-), and in the process release H^+ ions.
- Plant root activity: Plants take up nutrients in the form of ions (NO_3^-, NH_4^+, Ca^{2+}, $H_2PO_4^-$, etc.), and often, they take up more cations than anions. However plants must maintain a neutral charge in their roots. In order to compensate for the extra positive charge, they will release H^+ ions from the root. Some plants will also exude organic acids into the soil to acidify the zone around their roots to help solubilize metal nutrients that are insoluble at neutral pH, such as iron (Fe).
- Weathering of minerals: Both primary and secondary minerals that compose soil contain Al. As these minerals weather, some components such as Mg, Ca, and K, are taken up by plants, others such as Si are leached from the soil, but due to chemical properties, Fe and Al remain in the soil profile. Highly weathered soils are often characterized by having high concentrations of Fe and Al oxides.
- Acid Rain: When atmospheric water reacts with sulphur and nitrogen compounds that result from industrial processes, the result can be the formation of sulphuric and nitric acid in rainwater. However the amount of acidity that is deposited in rainwater is much less, on average, than that created through agricultural activities.
- Mine Spoil: Severely acidic conditions can form in soils near mine spoils due to the oxidation of pyrite.
- Decomposition of organic matter by micro organisms releases CO_2 which when mixed with soil water can form carbonic acid (H_2CO_3)

Sources of Basicity

Basic soils have a high saturation of base cations (K^+, Ca^{2+}, Mg^{2+} and Na^+). This is due to an accumulation of soluble salts which are classified as either saline soil, sodic soil, saline-sodic soil or alkaline soil. All saline and sodic soils have high salt concentrations, with saline soils being dominated by calcium and magnesium salts and sodic soils being dominated by sodium. Alkaline soils are characterized by the presence of carbonates. Soil in areas with limestone near the surface are alkaline from the calcium carbonate in limestone constantly mixing with the soil. Groundwater sources in these areas contain dissolved limestone.

Effect of Soil pH on Plant Growth

Acid Affected Soils: Plants grown in acid soils can experience a variety of symptoms including aluminium (Al), hydrogen (H), and/ or manganese (Mn) toxicity, as well as potential nutrient deficiencies of calcium (Ca) and magnesium (Mg).

Aluminium toxicity is the most widespread problem in acid soils. Aluminium is present in all soils, but dissolved Al^{3+} is toxic to plants; Al^{3+} is most soluble at low pH, above pH 5.2 little aluminium is in soluble form in most soils. Aluminium is not a plant nutrient, and as such, is not actively taken up by the plants, but enters plant roots passively through osmosis. Aluminium damages roots in several ways: In root tips and Aluminium interferes with the uptake of Calcium, an essential nutrient, as well as bind with phosphate and interfere with production of ATP and DNA, both of which contain phosphate. Aluminium can also restrict cell wall expansion causing roots to become stunted.

Below pH 4, H^+ ions themselves damage root cell membranes.

In soils with high content of manganese (Mn) containing minerals, Manganese toxicity can become a problem at pH 5.6 and below. Manganese, like aluminium becomes increasingly more soluble as pH drops, and Manganese toxicity symptoms can be seen at pH's below 5.6. Mn is an essential plant nutrient, so plants transport manganese into leaves. Classic symptoms of manganese toxicity are crinkling or cupping of leaves.

Nutrient Availability in Relation to Soil pH

Nutrients needed in large amounts by plants are referred to as macronutrients and include nitrogen (N), phosphorus (P), potassium (K), calcium (Ca), magnesium (Mg) and sulphur (S). Elements that plants need in trace amounts are called trace nutrients or micronutrients. Trace nutrients are not major components of plant tissue but are essential for growth. They include iron (Fe), manganese (Mn), zinc (Zn), copper (Cu), cobalt (Co), molybdenum (Mo), and boron (B). Both macronutrient and micronutrient availability are affected by soil pH. In slightly to moderately alkaline soils, molybdenum and macronutrient (except for phosphorus) availability is increased, but P, Fe, Mn, Zn Cu, and Co levels are reduced and may adversely affect plant growth. In acidic soils, micronutrient availability (except for Mo and Bo) is increased. Nitrogen is supplied as ammonium (NH_4) or nitrate (NO_3) in fertilizer amendments, and dissolved N will have the highest concentrations in soil with pH 6–8. Concentrations of available N are less sensitive to pH than concentration of available P. In order for P to be available for plants, soil pH needs to be in the range 6.0 and 7.5. If pH is lower than 6, P starts forming insoluble compounds with iron (Fe) and aluminium (Al) and if pH is higher than 7.5 P starts forming insoluble compounds with calcium (Ca).

Most nutrient deficiencies can be avoided between a pH range of 5.5 to 6.5, provided that soil minerals and organic matter contain the essential nutrients to begin with.

Determining pH

Methods of determining pH include:

- Observation of soil profile: Certain profile characteristics can be indicators of either acid, saline, or sodic conditions. Strongly acidic soils often have poor incorporation of the organic surface layer with the underlying mineral layer. The mineral horizons are distinctively layered in many cases, with a pale eluvial (E) horizon beneath the organic surface; this E is underlain by a darker B horizon in a classic podzol horizon sequence. This is a very rough gauge of acidity as there is no correlation between thickness of the E and soil pH. E horizons a few feet thick in Florida usually have pH just above 5 (merely "strongly acid") while E horizons a few inches thick in New England are "extremely acid" with pH readings of 4.5 or below. In the southern Blue Ridge Mountains there are "ultra acid" soils, pH below 3.5, which have no E horizon. Presence of a caliche layer indicates the presence of calcium carbonates, which are present in alkaline conditions. Also, columnar structure can be an indicator of sodic condition.
- Observation of predominant flora. Calcifuge plants (those that prefer an acidic soil) include *Erica*, *Rhododendron* and nearly all other Ericaceae species, many birch (*Betula*), foxglove (*Digitalis*), gorse (*Ulex* spp.), and Scots Pine (*Pinus sylvestris*).

Calcicole (lime loving) plants include ash trees (*Fraxinus* spp.), honeysuckle (*Lonicera*), *Buddleja*, dogwoods (*Cornus* spp.), lilac (*Syringa*) and *Clematis* species.

- Use of an inexpensive pH testing kit based on barium sulphate in powdered form, where in a small sample of soil is mixed with water which changes colour according to the acidity/alkalinity.
- Use of litmus paper. A small sample of soil is mixed with distilled water, into which a strip of litmus paper is inserted. If the soil is acidic the paper turns red, if alkaline, blue.
- Use of a commercially available electronic pH metre, in which a rod is inserted into moistened soil and measures the concentration of hydrogen ions.

Examples of Plant pH Preferences

- pH 4.5–5.0: Ericaceae (Azalea, Bilberry, Blueberry, Cranberry, Heather), Hydrangea for blue, (less acidic for pink), Liquidambar or Sweet Gum, Orchid, Pin Oak.
- pH 5.0–5.5: Boronia, Daphne, Ericaceae: (Camellia, Heather, Rhododendron), Ferns, Iris, Orchids, Parsley, Conifers (e.g., Pine), Poaceae: (Maize, Millet, Rye, Oat), Radish, Solanales: (Potato, Sweet Potato)
- pH 5.5–6.0: Asteraceae: (Aster, Endive), Brassicaceae: (Brussels sprout, Kohlrabi), Carrot, Cucurbitales: (Begonia, Chayote or Choko), Fabaceae: (Bean, Crimson Clover, Peanut, Soybean), Petunia, Rhubarb, Violet, most bulbs (Canna, Daffodil, Jonquil), Larkspur, Primrose
- pH 6.0–6.5 Antirrhinum or Snapdragon, Brassicaceae: (Broccoli, Cabbage, Candytuft, Cauliflower, Turnip, Wallflower), Cucurbitaceae: (Cucumber, Pumpkin, Squash), Fabaceae: (Pea, Red Clover, White Clover), Gladiolus, Iceland Poppy, Rosales: (Cannabis, Rose, Strawberry), Solanaceae: (Eggplant or Aubergine, Tomato), Sweet corn, Violaceae: (Pansy, Viola), Zinnia or Zinnea.
- pH 6.5–7.0: Amaranthaceae: (Beet, Spinach), Apiaceae: (Celery, Parsnip), Asparagales: (Asparagus, Onion), Asteraceae: (Chrysanthemum, Dahlia, Lettuce), Carnation, Fabaceae: (Alfalfa, Sweet pea), Melons, Stock, Tulip

Changing Soil pH

Increasing pH of Acidic Soil: The most common amendment to increase soil pH is lime ($CaCO_3$ or $MgCO_3$), usually in the form of finely ground agricultural lime. The amount of lime needed to change pH is determined by the mesh size of the lime (how finely it is ground)and the buffering capacity of the soil. A high mesh size (60–100) indicates a finely ground lime, that will react quickly with soil acidity. Buffering capacity of soils is a function of a soils cation exchange capacity, which is in turn determined by the clay content of the soil, the type of clay and the amount of organic matter present. Soils with high clay content, particularly shrink–swell clay, will have a higher buffering capacity than soils with little clay. Soils with high organic matter will also have a higher buffering capacity than those with low organic matter. Soils with high buffering capacity require a greater amount of lime to be added than a soil with a lower buffering capacity for the same incremental change in pH.

Other amendments that can be used to increase the pH of soil include wood ash, industrial CaO (burnt lime), and oyster shells. White firewood ash includes metal salts which are important for processes requiring ions such as Na^+ (sodium), K^+ (potassium), $Ca2^+$ (calcium), which may or may not be good for the select flora, but decreases the acidic quality of soil.

These products increase the pH of soils through the reaction of CO_3^{2-} with H^+ to produce CO_2 and H_2O. Calcium silicate neutralizes active acidity in the soil by removing free hydrogen ions, thereby increasing pH. As its silicate anion captures H^+ ions (raising the pH), it forms monosilicic acid (H_4SiO_4), a neutral solute.

Decreasing pH of Basic Soil

- Iron sulphates or aluminium sulphate as well as elemental sulphur (S) reduce pH through the formation of sulphuric acid.
- Urea, urea phosphate, ammonium nitrate, ammonium phosphates, ammonium sulphate and monopotassium phosphate fertilizers.
- organic matter in the form of plant litter, compost, and manure will decrease soil pH through the decomposition process. Certain acid organic matter such as pine needles, pine sawdust and acid peat are effective at reducing pH.

Acid Soils: Problems and Reclamation

Soil reaction is a very important chemical property of the soil. It is measured in terms of pH. The soils having pH 7 are neutral in reaction and those below or above 7 are acidic or saline (alkaline) respectively. In neutral soils, satisfactory plant growth can be expected, provided all other agronomic practices are followed properly. But neither acid soils nor saline soils are suitable for most of the crops except a few tolerant ones. Acidity or salinity acts as a limiting factor in crop production.

When the concentration of some elements which are naturally acidic in reaction, goes too high in the soil, soils become acid and their pH goes down below 7. These elements are basically hydrogen (H), aluminium (AI) and Iron (Fe). In fact, last two elements are not directly responsible for soil acidity but they have a tremendous potential to generate hydrogen ions (H+) in the soil which are direct cause of acidity (ion is the minutes possible and electrically charged form of an element). The more the number of hydrogen ions in a solution, the greater will be its acidity and the lesser pH count. This relation holds good in soil solution also. Therefore, the soils having hydrogen ions concentration above a reasonable limit or pH below 7 are termed as acid soils.

Soil acidity is of two types namely active acidity and reserve acidity. When hydrogen ions are present in the soil solution, this condition is known as active acidity. When the hydrogen ions are absorbed in the soil particles, this condition gives rise to reserve acidity. Reserved acidity is also called exchangeable acidity or potential acidity. The groups of hydrogen ions are always in equilibrium, the absorbed ions being greater in number than those present in the soil solution.

Genesis or Formation of Acid Soils

There are many factors responsible for formation of acidic soils. These can be discussed as under.

Climate

Acid soils are mostly found in the areas of high rainfall. In these regions, evaporation of water from the soil surface is less than precipitation. Excess amount of water moves downward in the soil and meets with the ground water table. This downward moving water carries a large amount of soluble salts from the soil surface. A reasonable amount of these salts is a must for maintaining neutral reaction of surface soil. In absence of these salts. Surface soil is, therefore, rendered acidic in reaction. Some compounds of iron and aluminium are resistant to leaching and, therefore, left on the surface. These compounds are acidic in nature and impart acidity to the surface soil (eg. laterite soils). In India, it is believed that the regions with acid soils must not receive more than 750 mm annual rainfall. The regions with annual rainfall of 1350 mm have acid soils with pH value 5 or even lower. In hilly regions where the loss of water through evaporation is very low due to low temperature, the conditions for the development of acid soils are very favourable although the rainfall is not so heavy.

Vegetation

In the regions of very low temperature the acid soils can develop easily. The foliage of coniferous trees which are most likely to grow in these regions is devoid of alkali elements. When the leaf litter of coniferous trees on ground is degraded, organic acids are released which gradually make the soil acidic. Vegetation of coastal zones and marshy places facilitates development of acidity. The places where excessive vegetation is accompanied by heavy rainfall and water pounding, soils are very acidic in reaction. In these places, plant residues are constantly submerged in water and they do not get enough air to decompose completely. This under-decomposed organic material keeps on piling up and soil becomes very acidic but at the same time very rich in organic matter.

Parent Material

Some young soils exhibit acidic nature even in the areas of medium rainfall and vegetation. These soils still reflect the nature of rocks from-which they have developed. Some rocks (parent material) are acidic in nature (e.g. igneous rocks). After weathering of these rocks, acidic constituents dominate the composition of soil. The rocks of soft and simple composition are especially suitable for development of acidic soils as they allow quick percolation of water through them and have less amount of alkaline elements and salts.

Topography

Sloppy places with good drainage conditions are often acidic in nature. On hill slopes, development of acid soils is easier. Acid soils do not develop generally in river basins. The plains with good drainage may also develop slight acidity in due course of time.

Human Interference

Excessive use of water or keeping the field submerged for a long time accompanied by improper drainage may lead to the development of acidity in the soil. Continuous use of acid forming fertilizers (eg.

ammonium sulphate, ammonium chloride, etc.) for years may also lead to considerable reduction in soil pH.

Effect of Soil Acidity on Plants

Except for a few crops like tea, coffee and potato, most of ! the crops do not find optimum conditions for their growth and Ii development The effects are more pronounced when the soil pH is below 6. Acidity affects the plants both directly and indirectly.

Direct Effects

1. Extremely high concentration of hydrogen ions just outside the plant roots has a toxic effect on roots.
2. The permeability of the plasma membranc surrounding root I hair is reduced. Consequently, the activity of root hair that absorb water and nutrients from the soil, is highly slackened.
3. Excessive presence of hydrogen ions in soil solution also results in increased absorption of $H+$ ions by the plant roots. This phenomenon disturbs acid-base balance inside the plant which virtually inhibits plant growth.
4. Numerous essential chemical reactions are carried out by the enzymes secreted by soil organisms as well as plant itself. These enzymes loose their effectiveness if hydrogen ion concentration goes too high.

Indirect Effects

1. Excess amount of hydrogen ions in the soil suppresses the availability of some essential elements such as calcium, magnesium and phosphorus, even if they are present in the soil in adequate amounts.
2. Some essential micronutrients such as iron, manganese, copper, etc. are required by the plants in very little amounts. They become highly soluble in acid soils and, their availability to the plant roots goes so high that they become toxic to the plants.
3. In most of the acid soils plants show deficiency symptoms of calcium and magnesium. This is because of the fact that acid soils are mostly deficient in these elements.
4. Some disease causing agents (especially fungi) flourish well in acid soils. Incidence of disease is, therefore, increased.
5. At the same time a large population of beneficial microorganisms suffers badly due to high concentration of hydrogen ions. This results in decreased soil fertility.

Classification of Acid Soils

For the convenience of study, acid soils of India are classified based on following factors:

Soil pH: pH is a good practical tool on which soils can be classified.

Table: *Add soil groups based on pH*

Sl.No.	*Soil type*	*pH range*
1	Slight acidic	6.6 – 6.1
2	Medium acidic	6.0 – 5.6
3	Strong acidic	5.5 – 5.1
4	Very strongly acidic	5.0 – 4.6
5	Extremely strongly acidic	4.5 or below

Organic Matter Content

Based on the amount of organic matter present, acid soils are placed under two categories:

Acid Mineral Soils

In these soils organic matter is less than 20 per cent Two types of soil namely podzol and laterite are put under this category. Podzol soils are mostly found in thickly forested hilly areas and they contain more organic matter than laterite soils. Decomposing organic residues produce a large amount of acids which along with rainwater remove salts from surface soil. Surface soil is thereby left acidic. Laterite soils are found in plains where temperature fluctuation in different seasons is high. Presence of porous, honeycomb-like reddish soil blocks below plough layer is a distinguishing character of laterite soils. These soil blocks are very hard and rich in iron and aluminium. Cultivated laterite soils are mostly reddish in colour.

Acid Organic Soils

These soils contain more than 20% organic matter. In the areas where heavy vegetation is accompanied with high rainfall and water logging conditions, acid organic soils are formed. These soils are again divided into two groups.

Peat Soils

In these soils a large amount of organic matter is present, but it is not fully decomposed. Due to high moisture content, adequate air is not available for complete decomposition of organic matter. These soils are generally submerged with water during monsoon season. The soils are black, heavy and highly acidic in nature

Muck Soils

In these soils, organic matter is found in highly degraded form. Unlike peat soils, the source of organic matter can not be identified in muck soils. These soils are found in Kerala, Kashmir, Himachal Pradesh and some parts of Assam.

Extent and Distribution of Acid Soils in India

A large chunk of Indian soils is reported to be acidic. Based on general estimations, nearly 25 million hectares of land is having pH below 5.5 and 23 million hectares fall under the pH range of 5.6-6.5. In all those tracts of the country where rainfall and temperature are high, acid soils are predominantly found. North-east region has the largest stretches of acid soils followed by neighbouring states of West Bengal, Bihar and Orissa. In the coastal region of I Kerala, not only high rainfall and temperature have contributed to the development of acid soils but rich organic deposits resulting in formation of peat and muck have also been found responsible for the extension of acid soil area in the state. Parts of Tamil Nadu, Andhra Pradesh, Karnataka, Madhya Pradesh, Uttar Pradesh, Bihar, Himachal Pradesh and Jammu & Kashmir also have acid soils.

Table: *Acid soils of India*

Sl. No	*Soil Group*	*PH range*	*characteristic*	*Area (m.ha)*	*States*
1	Laterite soils	4.8-7.0	Subsoil has a compact to vestibular mass containing compounds of sesquioxides	12.65	Karnataka, M.P., WB,MR, Kerala, Bihar
2	Laterite and lateritic red soils	5.0-7.0	Top soil deep and gravelly. Subsoil immature. Clay and iron compounds leach down to B-horizon	11.80	Kerala, Orissa,Assam, Santhal Parganas, and Singhbhum in Bihar
3	Mixed red and black soils	5.5 – 6.5	Free of lime concrete is found on surface. Colour varies according to amount of moisture and iron compound	23.66	Karnataka, Bihar, M.P. U.P
4	Ferruginous red soils	5.0-6.8	Sesquioxide is deposited in spots, limi-concretions are found, stony, some times gravelly	8.9	Tamil nadu, Parts of Karnataka, South East Maharastra, A.P. Goa, Orissa and Bihar.
5	Podozolic and brown forest soils	4.5-6.5	Formed in temperate zones under coniferous vegetation; Humus, iron aluminium leach down and deposit in B-horizon of soils.	22.47	J & K, UP, Assam, WB, Orissa, MP, HP.
6	Foot hill soil	4.5-4.6	Formed by alluvial particles brought down by rivers from hills, mostly sandy.	8.0	J & K,UP, Bihar, W.B.
7	Peat soils & muck	3.5-4.5	Un-decomposed and decomposed organic matter (more than 20%)	0.2	Kerala, J & K, H.P, UP

Reclamation of Acid Soils

Acidity in the soil is caused by excessive concentration of hydrogen ions. Suppressing hydrogen ions by applying lime is the best possible remedy of soil acidity. More than 90% of the lime used for this purpose in our country is generally in the form of calcium carbonate or lime stone (CaCO3). Besides, many other' liming materials are also used for the reclamation of acid soils according to their local availability.

Liming Materials

As stated above, calcium carbonate is the most widely used and the cheapest available liming material in our country. Depending upon availability many other amendments can also be used. These are briefly described below.

Calcium Oxide (CaO)

This is also known as un-slaked lime or burnt limestone or quick lime. It is a white powder and difficult to handle as it is caustic and explosive.

Calcium Hydroxide (Ca(OH)z

This is commonly known as slaked lime or building lime. It is also a white powdery substance with some loose lumps and is unpleasant to handle. This is prepared by mixing water with calcium oxide under controlled conditions. Much heat is generated in the process. Later, the material is dried and packed for marketing.

Dolomite [(Ca Mg (CO3)z]

Dolomite deposits are found in many places of India. Besides calcium, it also contains nearly equal amount of magnesium which is an essential plant nutrient Acid soils are often deficient in this nutrient.

Marl [CaCO3]

Marl is a soft, unconsolidated form of calcium carbonate.

Slag

Slags are the by-products of iron and steel industries. They are very rich in calcium as well as phosphorus. Main available slags are basic slag, electric furnace slag and blast furnace slag.

Lime Sludge

Some industrial wastes such as paper mill sludge are very rich in lime and after drying and processing they can be used as liming material. They are available free of cost in the areas where paper mills are located.

The required amount of various amendments is expressed in terms of calcium carbonate equivalents (CCE). The CCE means amount of calcium carbonate which can neutralise as much acidity as 100 kg of a liming material in question can neutralise. 100 kg of CaO can neutralize as much acidity as 179 kg of calcium carbonate and vice versa.

Table: *Calcium Carbonate (CaCO,) equivalent of some important liming materials*

Sl.No.	Liming material (100 kg)	Calcium carbonate equivalent (kg)
1	Calcium oxide (CaO)	179
2	Calcium hydroxide, (burnt lime)	139
3	Dolomite	109
4	Limestone (calcium carbonate)	100
5	Marl	70-90
6	Blast furnace slag	70-90
7	Electric furnace slag	65-80
8	Basic slag	60-70

If used on CaCO3 equivalent basis, all the liming materials have the same impact on crop yield. However, limestone is found to be superior to all other materials when tested in wide range of conditions. If available freely, basic slag is preferred to limestone in phosphorus deficient areas. Likewise for magnesium deficient areas, dolomite is superior. For acid soils of Kerala, burnt lime has been found very suitable.

Amount of Lime Required

Soil testing for pH is essential for finding out approximately correct dose of liming material. From an acre of land 6-8 samples should be dug out randomly from a depth of about 15-20 cm. All these samples should be mixed thoroughly and one representative sample weighing about half kg should be taken to the nearest soil testing laboratory. The report given by the testing laboratory provides pH value of the soil and recommended dose of pure calcium carbonate to be used in an acre or hectare of land. Even if amount of liming material is not given, it can be found out from the table taking the help of soil testing report.

Table: *Lime requirement to bring the soil to Indicated pH*

pH of soilbuffer suspension	***Lime required to bring the soil to indicated pH (in tones per acre of pure calcium carbonate i.e.$CaCO_3$)***		
	$pH_{6.0}$	$pH_{6.4}$	$pH_{6.8}$
6.7	1.0	1.2	1.4
6.6	1.4	1.7	1.9
6.5	1.8	2.2	2.5
6.4	2.3	2.7	3.1
6.3	2.7	3.2	3.7
6.2	3.7	3.7	4.2
6.1	3.5	4.2	4.8
6.0	3.9	4.7	5.4
5.9	4.4	5.2	6.0
5.8	4.8	5.7	6.5
5.7	5.2	6.2	7.1
5.6	5.6	6.7	7.7
5.5	6.0	7.2	8.3
5.4	6.5	7.7	8.9
5.3	6.9	8.2	9.4
5.2	7.4	8.6	10.0
5.1	7.8	9.1	10.6
5.0	8.2	9.6	11.2
4.9	8.6	10.1	11.8
4.8	8.1	10.6	12.4

The amount of the liming material greatly depends upon the soil texture. The smaller the soil particles, the greater will be the site exposed for hydrogen ions to stick on. Neutralizing the hydrogen ions present in the soil solution (active acidity) is not enough. The amount of amendment should be enough to neutralize hydrogen ions swarming around the soil particles also (reserve acidity). Soil testing for lime recommendation must take this point into account. Reserve acidity is sometimes thousand times greater than active acidity In clay soils, reserve acidity is very high, whereas in sandy soils it is comparatively low. The amount of liming material required to neutralize total acidity (active + reserve), therefore, varies greatly according to soil texture.

Table: *Calcium Carbonate (limestone) needed to increase the pH of a soil textural class by one unit.*

Sl. No.	*Soil texture*	*Lime tones/ha*
1	Sand	2.50
2	Sandy loam	3.75
3	Loam	5.00
4	Silt loam	6.25
5	Clay loam	7.50
6	Clay	8.25

Note: In the ready reckoner, amounts of calcium oxide (CaO) have been given which is not a commonly used amendment. To convert the amount of calcium oxide into limestone (calcium carbonate) multiply it with 1.79. It should be noted that amounts are given in pounds per acre.

Fineness of Liming Material

For quick results, liming material should be properly ground before application to the field. Application of coarse material may mislead us even if applied in appropriate amount Coarse particles take unusually long time to dissolve in water and interact with hydrogen ions in the soil and, therefore, quick response in terms of increased soil pH is not achieved. Particles of the size which can pass through 80-100 mesh sieve (a sieve having 80-100 holes per square centimeter area) are quite fine and show hundred per cent efficiency.

Application of Lime

Lime is not mobile in the soil. Therefore, it should be spread in the field with maximum possible uniformity and worked well into the soil. This can be done during the preparation of field. Soil should be sufficiently moist at the time of liming or a light irrigation may be given to the field after mixing the lime. If slaked lime is used, seeds should be sown in the field at least 3-4 weeks after liming. To maintain desirable soil reaction in the humid regions, liming at every 3-5 years interval is recommended.

Advantages of Liming

Advantages of liming are three fold viz. physical, chemical and biological. They are described below.

Physical Effects

Liming improves physical condition of heavy soils; they become granular in structure and their water holding capacity is improved.

Liming also encourage the decomposition of organic matter and consequently, there is greater production of organic colloids. The Ca-humus so produced is believed to be an effective cementing agent in binding the soil particles. Liming also prevents soil erosion because soils which have received liming treatment support good plant growth.

Chemical Effects

Among the principal chemical properties influenced by liming is the reduction of H ions in the colloidal complex. It increases the availability of almost all the nutrients such as nitrogen, phosphorus, potassium, calcium, magnesium, boron, zinc, copper and molybdenum, and reduces the toxicity caused by soluble iron, aluminium and manganese.

Biological Effects

One of the outstanding biological effects of liming is to encourage the microbial activity of the soil. By raising the soil pH, it make the soil more congenial for a number of micro organisms. Nitrifying and nitrogen fixing bacteria, both symbiotic and non symbiotic are stimulated by the addition of lime to an acid soil. Lime also brings about a more rapid decomposition of organic manure, both native and added, as a result of improved microbial activity. This further increases the availability of nitrogen phosphorus and sulphur.

Adverse Effects or Over Liming

Lime may have adverse effects especially if the soil is over limed. If lime is added to an acid soil until the pH goes above 7, it produces certain deleterious effects on the crop. The harmful effect is more pronounced on soils that are not well buffered such as sandy soils and soils poor in organic manure. The main effect of over liming is to reduce the availability of some of the essential nutrients, both major and minor, such as iron, manganese, boron, copper, zinc, phosphorus, potassium, etc., and thus bring about nutritional deficiency. Excess lime also interferes with the absorption of certain elements like potassium, phosphorus, boron, etc. by plants thus hindering their utilization. The very rapid decomposition of organic matter in soils of arid of semi arid regions is also attributed to the accumulation of excess lime in these soils.

Crops Tolerant to Acidity

Though most plant species of agricultural importance can grow satisfactorily in culture solutions having a wide range of pH, their behaviour in soil is very different. Most crop plants grow well when the soil pH is between 5.5 and 7.5, a fairly wide range even under field condition. Serious trouble develops when the pH drops below 5.0. A large proportion of crop plants can tolerate slight to moderate acidity, a few can tolerate even a fairly strong acidity. Crops which can be successfully grown in acid soils are listed in the following table.

Table: *Add tolerant crops*

Sl. No.	*Slightly tolerant*	*Medium tolerant*	*Very tolerant*
1	Berseem	Maize	Mustard
2	Sugarcane	Potato	Buck wheat
3	Cauliflower	Wheat	Coffee
4	French bean	Soya bean	Rubber
5	Cabbage	Barley	Tea
6	Watermelon	Oats	—
7	Lucerne	Rice	—

Plant Nutrients

Like human beings, plants too need the right kind and amount of food for their growth and development. But unlike man, plants cannot move around to collect their food. They have to synthesize their own food. To synthesize their food, plants require nutrients in addition to light, air, heat and water. Plant nutrients are therefore, defined as the elements required by the plants to synthesize their food and maintain physiological activities for plant growth and development. Most of these nutrient elements are taken by the plants from soil and some from the air. These nutrients are called essential elements.

Essential Elements

What are the essential elements? Chemical analysis of a plant body has revealed the presence of some sixty elements in palnt-body. However, all are not essential for plant growth and development. Only seventeen plant nutrients are so far known to be essential for plants. To decide whether an element is essential or not the following three criteria have been laid down by the scientists.

1. A deficiency of such element makes it impossible for the plant to complete its life cycle, i.e. vegetative and reproductive stages of plant life.
2. The deficiency symptoms of such element are very specific and can be corrected by supplying only that element.

3. The element is directly involved in the nutrition and other physiological processes of the plant, quite apart from its possible effect in correcting some micro-biological or chemical conditions of the soil. Its effect is quite independent and well defined.

For an element to be essential, it must follow at least one or more of the criteria given above. But it should also be kept in mind that some elements are essential for all plants but all elements are not essential for all plants.

Calssification of Plant Nutrients

Although there is no well defined scale to classify plant nutrients, they are broadly classified as (a) macro nutrients and (b) micro-nutrients.

Macro-nutrients

Nutrient elements which are required by the plants relatively in very large amounts are grouped as macronutrients. Nine elements are placed under this group which are again classified into the following groups.

Skeleton Nutrients

Carbon (C), hydrogen (H) and Oxygen (O) form the major portion of plant body. But they are not a problem for plants body. But they are not a problem for plants since they can easily reach inside the plant through air and water. These elements are the major constituents of the plant tissues (a group of cells) which provide mechanical strength to plants. They are, therefore, called skeleton elements.

Primary Nutrients

Primary nutrients are required by the plants in a large amount and soils are very often deficient in them. Nitrogen (N), phosphorus (P), and potassium (K) are placed under this group. These nutrient elements are applied in the soil artificially through fertilizers and therefore, they are also called fertilizer elements.

Secondary Nutrients

In the order of nutritional importance to plants secondary nutrients come next to primary nutrients. Calcium (Ca), magnesium (Mg) and sulphur (S) are grouped under this category.

Micro-nutrients

Though essential, the micro-nutrients are required by the plants in very minute amounts or traces. For example, only one spoonful of molybdenum in one hectare is enough to meet the plant need, while many kilograms of nitrogen may not enough to meet the nutrient need of the plant. Eight elements, namely, iron (Fe), manganese (Mn), zinc (Zn), copper (Cu), boron (B), molybdenum (Mo), chlorine (Cl) and cobalt (Co) have identified as micronutrients. Table provides the classification of the plant nutrients.

Table: *Classification of plant nutrients.*

Macro-nutrients			***Micro-nutrients***
Skeleton nutrients	***Primary nutrients***	***Secondary nutrients***	
Carbon Hydrogen Oxygen	Nitrogen Phosphorus Potassium	Calcium Magnesium Sulphur	Iron Manganese Copper Zinc Boron Molybdenum Chlorine Cobalt
From air & water	From the soil		

Function of Plant Nutrients

Various functions of plant nutrient are described as follows.

Carbon, Hydrogen and Oxygen

Carbon, hydrogen and oxygen form the skeleton of the plant body. They are the major and essential components of almost all the chemical substances found in the plant body.

Nitrogen

Nitrogen amounts to 1 to 4% of dry weight of the plant. It is taken up from the soil in the form of nitrate (NO_3) and ammonium (NH_4) and combines with carbohydrates formed by photosynthetic process to form protein. Nitrogen is an integral constituent of protein. Protein is used by the plants as a raw material in the formation of cells and tissues. Protein is also called a body building substance. A good amount of nitrogen supply to the plant is important for the uptake of other nutrient. Excessive nitrogen supply to plant is harmful as it creates excessive vegetative growth at the cost of grain formation. Excessive vegetative growth also makes the crop more susceptible to lodging and pest attack.

Phosphorus

Phosphorus accounts for 0.1 to 0.4 % of dry matter in the plant. It plays a key role in the transfer of energy inside the plant. Thus it is essential for photosynthesis and other physiological processes taking place in the plant. It is indispensable for cell division and for the development of tissue which form the growing points of the plants (stem top and root top). Phosphorus also helps in root growth. It is particularly useful for leguminous crops as it increases the activity of nitrogen fixing bacterial living inside the nodules located on roots of legume plants. Phosphorus helps the formation of seeds and fruits, particularly in legumes.

Potassium

Potassium forms 1-4% of the dry matter of the plant and remains dissolved in cell sap inside the plant. The function of potassium is manifold. It activates more than 60 enzymes (Chemical substances which carry out many life governing processes) and plays a vital role in carbohydrates (sugars) and proteins from leaves to roots. It imparts vigor and resistance to plants against attack of pests. Some crops such as tomato, cloves, lucern, beans, tuber and root crops and oil seed crops are more responsive to potassium than to other nutrients.

Calcium

Clacium content of plant dry matter ranges from 0.5 to 3 %. It is essential for root growth and serves as constituent of cell-wall (cover) material. Most soils contain an abundance of available calcium. The aim of calcium application is usually to correct soil acidity rather than to increase the supply of calcium as a plant nutrient. Application of calcium in acid soils increases availability of other nutrients such as phosphorus, nitrogen and molybdenum. Excess of calcium in calcareous soils decreases the uptake of potassium and magnesium.

Magnesium

Nearly 15-20% of the magnesium contained in the plant is located in the green parts and is the central constituent of chlorophyll (the green pigment of the leaves), which carries out photosynthesis process in the presence of sunlight. Magnesium is also needed in enzymic reactions related to the energy transfer of the plant. It maintains the dark-green colour of leaves and enhances the uptake of other materials particularly nitrogen and phosphorus. It plays an important role in the movement of phosphorus and formation of oil and fat.

Sulphur

Sulphur is also an essential constituent of many types of protein. It helpes in the formation of fats and oils and accelerates root growth.

In legumes, activity of rhizobium (nitrogen supplying bacteria) is increased by sulphur application. Oil contents are also found to increase in oiseed crops if sulphur is supplied to the plants in optimum amount.

Micronutrients

Most of the micronutrients are either constituents of activators of enzymes. Enzymes, as already said. Are very important substances which carry out numerous physiological processes inside the plant. Iron and manganese although are not constituent of chlorophyll by forming enzymes. Molybedenum and cobalt, in the same way, affect nodule formation in leguminous crops. Other micronutrients also influence several vital processes taking place in the plant.

Deficiency Symptoms of Plant Nutrients

It is difficult to identify the deficiency enzymes of a particular type of plant nutrient because the symptoms created by two elements or even more can be very similar. Yet the knowledge of deficiency symptoms of individual nutrient coupled with knowledge of soil factors and the experience of farmer himself can provide greater accuracy in identifying a deficient nutrient. These symptoms can be studied nutrient wise as follows.

Carbon, Hydrogen & Oxygen

Plants seldom face the scarcity of these three elements, viz., carbon, hydrogen and oxygen, as they are available from air and water. However, oxygen can be deficient if the moisture is excessive and the soil very compact. Oxygen deficiency results in extremely stunted plant growth.

Nitrogen

The following are the most common symptoms of the deficiency of nitrogen.

1. Plants go light-green in colour, the lower leaves turn yellow and in some cases they quickly start drying up as if suffering from shortage of water.
2. The growth is stunted and stems are dwarfed.
3. In cereals, tilling is restricted and stem becomes spindle shaped.
4. In potato, at the later stages of growth, the margins of lower leaflets loose their green colour and become pale-yellow.
5. In cotton the leaf size is reduced, colour chage to yellow or brown and die. Plants produce fewer lateral branches, especially the fruiting branches.

6. In citrus, the leaf shedding is heavy. Leaves become thin, small, fragile and light in colour.
7. In deciduous fruit trees, mature leaves are discoloured from base to tip. Under prolonged deficiency, twigs become hard but thinner.
8. In vegetables, there is retarded growth with leaf chlorosis (disappearance of green pigment chlorophyll of leaves). The branches are slender, fibrous and hard.

Phophorus

The deficiency symptoms of phosphorus are the following.

1. Generally, the plant appears dark green in colour but lower leaves may turn yellow and dry up.
2. Incorn, leaves and stems show a tendency to become purplish; young plants become stunt and dark green in colour.
3. Wheat, barley and rice have dark green colour, often showing purplish tinge on leaves, often showing purplish tinge on leaves, and retarded growth.
4. In Potato, at the early stages, plants have stunted, spindly growth. The tubers have rustry frown lesions in the form of isolated flecks.
5. The cotton plants have dark green colour, leaves and stems are small and the bolls mature late.
6. Small leaves of legumes are curled upwards. The plants are stunted and spindly.
7. In citrus the growth is reduced. The older leaves loose their deep green colour and lusture and develop faded green or bronze colur. Dead spots appear on the leaves and expand further.
8. In vegetables although, the growth is retarded, the leaves do not show the symptoms of chlorosis.
9. In deciduous fruit trees, young leaves develops reddish purple colour and the new twigs are slender.
10. In many crops, the lower surface of leaves develops reddish purple colour. The stem is slender and woody. They bear small dark green leaves.

Potassium

The deficiency symptoms of potassium are the following.

1. The margins of leaves turn brownish and dry up, stem remains slender.

2. In tobacco, small spots of dead tissues appear between the veins and at leaf tips and margins which are tucked or cupped up.
3. In maize, at the young stage, the edges and tips become dry and appear scorched or fired. At a later stage in well grown plants the leaves are streaked with yellow and yellowish green colour and the margins dry up and get scorched. Similar symptoms are shown by oat, wheat and barley.
4. In potato, plant growth is retarded, the leaf size is reduced, leaves become crinkled and curl downward. The older leaves become yellowish and develop a bronze colour at the tips which advances towards t he leaf base and die.
5. In cotton, collar rot occurs. First the leaves show yellowish green, and subsequently yellowish spots appear between the veins. The centre of these spots dies and numerous brown specks occur at the top, around the margin and between the veins. The leaf curls downwards before it becomes reddish brown and dries up.
6. Yellow mottling around the edges of leaves takes place in legumes. These areas soon dry up and die.
7. In citrus, excessive leaf shedding takes place at blossom stage. Young shoots shed before they become hardened.
8. In deciduous trees, necrosis occurs on leaves. The necrotic area very in size from very small dots patches. Leaves especially of peach becomes crinkled.
9. In vegetable crops in the older leaves bronze and yellowish brown colours are manifested near the margins. Specks develop along the veins of the leaves. Ultimately the tissue deteriorates and die.

Calcium

The deficiency symptoms of Calcium are the following.

1. Generally, the deficiency symptoms of calcium are localized in new leaves and in bud leaves of plants. In severe cases the bud at the stem top (terminal bud) dies.
2. In tobacco, the young leaves forming the terminal bud first become typically hooked and die back at tips and margins. The stalk finally dies.
3. In maize, the tips of the unfolding leaves secrete a sticky material and the leaf tips stick together when dry.

4. In potato, a light green band appears along the margins of the young leaves of the bud. The leaves often wrinkle.
5. In cotton, the petiole which connects leaf blade with tem, bends and later collapses resulting in leaf shedding.
6. In vegetables, the stems grow thick and woody and the new leaves become chlorotic.
7. In legumes, the nodules developed on roots are small and fewer in number.
8. In citrus the green colour fades along the edges of the leaf and it spreads to areas between veins.

Magnesium

The following are the deficiency symptoms of Magnesuium

1. Generally, the symptoms first appear on old leaves.
2. In tobacco, the lower leaves are chlorotic but do not show dead spot. The tips and margins of the leaf are turned or cupped upwards.
3. In maize, leaves slim and yellow streak develops between the parallel veins in the leaves. In acute deficiency, these strip-tissues dry up and die.
4. In small cereal, plants are stunted and leaves develop yellow green patches.
5. In potato the affected leaves are brittle.
6. In legumes, chlorosis begins at the tip and margins of the lowermost leaf and progresses between the veins towards the centre of the leaf.
7. In cotton, the lower leaves have purplish red colour with green veins.
8. In citrus trees, the fading of green colour from areas parallel to the midrib occurs and spreads from there. Base of the leaf usually remains green.
9. In leaf shedding fruit trees, necroses occurs as fawn coloured patches on most mature, large leaves. The affected leaves drop.

Sulphur

The deficiency symptoms of sulphur are the following.

1. Chlorosis in the young leaves of plants takes place.
2. In tobacco, whole leaf becomes light green in colour and terminal bud remains alive.

3. Chlorosis in potato develops slowly but the growth of plants is materially checked.
4. Cotton plants dwarf but young leaves do not show colour change.
5. Leaves of vegetables crops become yellowish green, thick and firm. The stems harden and some times become abnormally elongated and spindle shaped.
6. In legumes the younger leaves become pale green to yellow.

Boron

The deficiency symptoms of Boron are the following.

1. In wheat, minute chlorotic specks appear on older leaves, which later give an orange tint and enlarge to form irregular areas of bright orange yellow spots in midleaves.
2. In barley inter-venal chlorosis takes place. When the plants is about 7 weeks old. Older leaves develop dark brown necrotic spots along with tips and margins. Eventually the necrosis spreads.
3. Five to six week old plants of maize develop white spots on leaf blade. Size of the spots increases to form large irregular necrotic lesions. The growth of terminal leaves is often reduced, they become chlorotic, fail to unroll and exude some soluble salt. No cobs are formed under sevear deficiency.
4. In paddy, small white necrotic streaks develop on the youngest leaves and enlarge to form white irregular patches. Leaves often fail to unroll.
5. Mustard leaves become thick and leathery with purple pigmentation and deformation of younger ones. Terminal bud dies flower buds fail to open and plants do not fruit.
6. In groundnut the number and the size of leaves are reduced. Shoot growth is suppressed. Plant dry weight is reduced by 55-75 per cent. Roots turn brown and brittle.
7. Cucumber shows dark green colour of leaves in the beginning which later turns dirty brown yellow and chlorotic. Leaves are malformed and leathery.

Copper

The deficiency symptoms of copper are the following.

1. In wheat, leaves remain unrolled with the tips severely chlorotic. Chlorosis starts from the leaf tip and progresses towards the

base. Plants show bluish green appearances. In severe deficiency, tips dry become papery, withered and twisted. Plant remains bushy and no flowering occurs.

2. In tomato leaves curl and old leaves become flaccid and drop. Entire plant shows wilted appearance. Young leaves do not enlarge, go pale and necrotic. Plant develops very poorly with less branches. Root growth is checked. Flowering and fruiting is severely restricted.
3. In guar, necrotic spots develop on leaves and the leaves curl and the plant is affected severely.
4. In cotton, the leaves turn yellowish green with white spots. The upper portion of the plant is affect the most but the veins remain green.
5. In mango, leaves dry out occasionally. They are bigger than the normal size.

Iron

The deficiency symptoms of Iron are the following.

1. In wheat, green colour of leaves deteriorates. Symptoms appear first on young leaves. Mild to Strong chlorosis on intravenial areas of barley leaves takes place. The severely affected plant ceases to grow and collapses.
2. In maize, young leaves develop severe chlorosis within 3-4 weeks of sowing. The leaves may totally bleach.
3. In paddy, chlorosis of young leaves is followed by scorching and drying up of leaf tips.
4. In seed crops, root development is suppressed. Chlorosis occurs but veins and petiole remain green. Younger leaves are more susceptible to deficiency. Fruit setting is very poor and flowers shed before fruiting.

Manganese

The deficiency symptoms of Manganese are the following.

1. Bud leaves are chlorotic but terminal but remains alive.
2. In tobacco, the young chlorotic leaves develop dead tissues scattered over the leaf. Veins remain green.
3. In oats grey speck diseases has been found associated wit h manganese deficiency.
4. In potato, numerous small brown patches develop which eventually become more extensive.

5. in cotton, bud leaves become yellowish-grey while veins remain green terminal bud alive.
6. In ceareals the leaves turn brown or transparent.

Addition of Nutrients

Nutrients come to the soil basically through two waus: (a) natural means, such as rain, weathering of parent material, fixation of atmospheric nitrogen in soil and plant by micro-organism, and (b) artificial means which involve the application of fertilizers, organic manures like cattle during and urine, compost, biofertilizers and soil amendments like gypsum, sulphur and pesticides.

Removal of Nutrients

The major factors, which account for the removal of nutrients from the soil are (i) crop harvest, (ii) weeds (iii) soil erosion, (iv) gaseous form and (v) leaching.

Crop Harvest

The main nutrient loss from the soil is caused by the crop harvest. The average amount of nitrogen, phosphorus, potassium, calcium, magnesium and sulphur removed from the soil by certain important crops.

Table: *Average Removal of Plant Nutrients by certain Crops*

Sl. No.	*Types of Crops Experimented*	*Economic Yield (Qt/ha)*	*Total Removal of Nutrients (Kg/hecatare)*					
			Nitrogen	*Phosphorus*	*Potassium*	*Calcium*	*Magnesium*	*Sulphur*
1.	Rice	30	84	14	89	21	9	9
2	Wheat	30	125	22	92	16	14	14
3	Sugarcane	880	180	26	270	132	--	25
4.	Onion	370	133	22	177	16	18	34
5.	Tomato	410	84	21	185	31	8	28
6.	Soyabean	25	125	43	101	35	19	22
7.	Groundnut	20	170	30	110	39	20	15
8.	Mustard	15	83	17	71	63	13	26

Weeds

If weeds are allowed to grow, large amounts of nutrients are drained out from the soil since they also like crop plants require all the macro and micro nutrients. Weeds are usually removed from the field and burnt. By this process large quantities of nutrients are lost from the soil. Removal of nutrients by weeds can be avoided by controlling weeds at the seedling stage or by composting and incorporating them into the soil.

Soil Erosion

Soil erosion also causes huge loss of nutrients from the soil along with the loss of soil itself. It is estimated that about 2.5 million tones of nitrogen, 3.8 million tones of phosphorus and 2.6 million tones of potash are lost annually from our country by soil erosion (Indian Farming, October 1986). This loss exceeds the amount of indigenously manufactured fertilizers in the year 1981-82. Erosion losses are predominant in agricultural land, soils which are light in texture and in areas where soil and water conservation measures are not adopted.

Gaseous Form

Loss of nitrogen from the soil also takes place by conversion of fixed form nitrogen in the soil into gaseous form. In gaseous form, nitrogen is liberated to atmosphere and con not be utilized by the plant. This conversion may be brought about by micro-organisms (denitrification) if soil aeration is poor, or it may be completely chemical (volatalization) if soil is very saline and alkaline in nature.

Leaching

In some light soils of wet regions, nutrients are leached down to the deeper layers and are no more available to the plant roots. The gain and loss of nutrients are affected by a number of factors like soil properties, type of crop and its variety, environmental conditions, etc. The productivity of a soil depends upon the balance between the gains and losses of nutrients. If cropping is continued without supplementing for the nutrient losses by crop harvest soil cannot sustain its productivity for a long time.

Absorption of Nutrients

Mechanism of absorption: It is interesting to know how nutrient elements manage to make their way into plant roots. In the soil nutrients remain either adhered on soil particles or dissolved in soil water. A nutrient must be water-soluble if it is to be absorbed by the plant roots. Plants absorb nutrients from soli particles, using water as a nutrients from soil particles, using water as a nutrient carrier between root hairs and soil particles. Root hairs are very fine hairy projects located on the lower part of the roots.

Plant roots (nutrients) soil-water (nutrients) soil particles

$\leftarrow\!\!-\!\!\rightarrow$ $\leftarrow\!\!-\!\!-\!\!-\!\!\rightarrow$

Soil water remains in direct contact with roots. Roots absorb nutrients from water leaving the water depleted of plants nutrients. Soil water. Then, extracts out nutrients are again replenished in the soil water. This dynamism is always going on in the soil, soil water and root system.

Forms in Which Nutrients are Absorbed: The nutrients are always absorbed in electrically charged form which is called ionic form. These ions may be positively charged (cation0 or negatively charged (anion). Nitrogen (Na0 is absorbed in nitrate (NO_3) and ammoniacal (NH_4) forms and calcium is absorbed in (Ca^{++}) form. If roots absorb an ion having positive charge, they have to absorb another ion also having negative charge of equal number. For example if root absorbs one ion of Ca^{++} it will have to absorb two nitrate (NO_3) ions so that electric charges inside the root is kept in balance.

Processes of Absorption: There are two processes of plant nutrient absorption; passive absorption and active absorption.

Passive Absorption: In passive absorption plant does not need to spend energy. If concentration of nutrients outside the root is more than that of the inside nutrients diffuse into the roots along with water. This process is called diffusion. If plant is in dire need for water and plenty of water is given to the plant at a time, some nutrients manage their way inside the plant roots along with rapidly moving water from outside. This process is called mass flow.

Active Absorption: In active absorption roots need energy generated from respiration process to absorb the nutrients from the soil. Using this energy roots cells develop suctions force and absorb nutrients from soil solution.

Maintenance of Plant Nutrients in the Soil

To meet the increasing demand for food grains it is necessary to sustain the fertility of the soil. Increased land use by growing more crops in a definite time span causes removal of plant nutrients is large amounts from the soil. To sustain the soil fertility in order to ensure successful cropping for a prolonged period, compensation to the soil for nutrient losses is essential. Some nutrient losses by soil erosion, leaching, denitrification, etc. but to manage the soils in an effective manner, these losses much not be neglected. Therefore both curative as well as preventive measures should be adopted to maintain optimum level of plant nutrients in the soil. The factors, which contribute to the maintenance of plant nutrients in soil are: (a) fertilizers, (b) organic manures, (c) soil amendments, (d) soil and water conservation, (e) drainage, (f) crop rotation, and (g) proper soil management practices.

Fertilizers

Fertilizer application is necessary to sustain soil productivity. Sometimes the soil may be already very rich in plant nutrients, but they are not so readily available to the plant as they are not in water soluble form. Fertilizers contain plant nutrients in readily available form. They can meet the pace of nutrient requirement by the plant. Yet the recovery of fertilizer nutrients by plants seldom exceeds 15-30 per cent. The remaining portion is either fixed with the soil particles or lost by various means from the soil. The fixed nutrients may again become available to subsequent crops when there is a stress upon nutrients. To supply micronutrients, micronutrient fertilizers and fertilizer mixtures have been developed but their use is restricted due to cost factors and technical difficulties in use.

Organic Manures

Organic manures like farm yard manure, compost, green manure oil cakes sewages; crop wastes etc. must be recycled into the soil if soil fertility is to be maintained. Organic manures release a large amount of plant nutrients when they undergo decomposition. They are the food for millions of useful micro-organisms inhabiting the soil. Unlike fertilizers, organic manures release micronutrients after decomposition. Further they provide a resistance capacity to the soil against excessive build up of any element and are essential for maintaining good soil physical conditions.

Soil Amendments

The amount and availability f plant nutrients in the soil is very much dependent upon soil reaction. In alkaline soils, availability of some micronutrients goes so low that plants may show the symptoms of micronutrient deficiency. Likewise in acidic soils most of the phosphorus is converted unto unavailable form and the nitrogen is coverted into gaseous form and released into the atmosphere. These types of losses are sometimes, exceedingly high depending upon the agro-climatic conditions. The amendments of these soils is necessary to ensure a good crop production. Material used as amendment of these soil is necessary to ensure a good crop production. Material used as amendments themselves contain some plant nutrients like lime, sulphur, etc., depending upon the type of amendment. Secondly these amendments have profound impact on the availability of other plant nutrients. Amendments push the soil reaction toward neutral where availability of most of the plant nutrients is just optimum.

Soil and Water Conservation

The most fertile top layer of the soil skimmed off through water and wind is called soil erosion. Erosion causes a great loss of nutrients from the soil but it is seldom noticed by the farmers especially when splash and sheet erosions are active. All attempts should be made to arrest the maximum amount of rainwater at the site of raining itself. Soil should not be allowed to be drained away along with run off water in scanty rainfall areas.

Drainage

In places where water logging takes place, aeration to the root zone is hampered or checked. Due to lack of oxygen, nitrogenous compounds (ammonium) of the soil are converted by micro organisms into gaseous form which is no more available to the plant roots. Nitrogen is therefore lost to the atmosphere in large amounts. Proper arrangement of drainage should be made to drain away excess amount of water so as to ensure good aeration into the soil. Thus the loss of nitrogen from the soil can be prevented and roots can be provided with enough amount of oxygen to carry out the energy generating respiration process.

Crop Rotation

A proper crop rotation must be followed to maintain soil fertility. Nutrient exhausting crops like sorghum, rice, sugarcane, wheat, jowar, bajar, maize etc. should be followed by nutrient replenishing crops like any of the legumes. Legume crops improve nitrogen content of the soil. Like wise, deep rooted crops should be followed by shallow rooted ones and so on.

Management Practices

All the agronomic practices must be given due consideration based on the type and condition of the soil. For example, if soil is sandy, all the dose of fertilizers should not be given at a time. Instead it should be given in two or three split doses. Likewise in sandy soil, shallow but frequent irrigations should be given. This way we can reduce the loss of water as well as nutrients from the sandy soil. Soils with low organic matter should receive more and frequent green and other organic manuring. Weeds should not be allowed to exhaust plant nutrients from the soil and whatever weeds grow should be uprooted and composted to make organic manures and recycles them back to the soil.

Nitrogen Cycle

The nitrogen cycle is the process by which nitrogen is converted between its various chemical forms. This transformation can be carried out through both biological and physical processes. Important processes in the nitrogen cycle include fixation, ammonification, nitrification, and denitrification. The majority of Earth's atmosphere (78%) is nitrogen, making it the largest pool of nitrogen. However, atmospheric nitrogen has limited availability for biological use, leading to a scarcity of usable nitrogen in many types of ecosystems. The nitrogen cycle is of particular interest to ecologists because nitrogen availability can affect the rate of key ecosystem processes, including primary production and decomposition. Human activities such as fossil fuel combustion, use of artificial nitrogen fertilizers, and release of nitrogen in wastewater have dramatically altered the global nitrogen cycle.

A 2011 study found that nitrogen from rocks may also be a significant source of nitrogen, that had not previously been included in most calculations and statistics. ogen is essential for many processes and is crucial for any life on Earth. It is a component in all amino acids, as incorporated into proteins, and is present in the bases that make up nucleic acids, such as RNA and DNA. In plants, much of the nitrogen is used in chlorophyll molecules, which are essential for photosynthesis and further growth. Although Earth's atmosphere is an abundant source of nitrogen, most is relatively unusable by plants. Chemical processing, or natural fixation (through processes such as bacterial conversion), are necessary to convert gaseous nitrogen into forms usable by living organisms, which makes nitrogen a crucial component of food production. The abundance or scarcity of this "fixed" form of nitrogen, (also known as reactive nitrogen), dictates how much food can be grown on a piece of land. The nitrogen cycle and the carbon cycle work together to maintain an ecosystem.

The Processes of the Nitrogen Cycle

Nitrogen is present in the environment in a wide variety of chemical forms including organic nitrogen, ammonium (NH_4^+), nitrite (NO_2^-), nitrate (NO_3^-), nitrous oxide (N_2O), nitric oxide (NO) or inorganic nitrogen gas (N_2). Organic nitrogen may be in the form of a living organism, humus or in the intermediate products of organic matter decomposition. The processes of the nitrogen cycle transform nitrogen from one form to another. Many of those processes are carried out by microbes, either in their effort to harvest energy or to accumulate nitrogen in a form needed for their growth.

Nitrogen Fixation

Nitrogen fixation is a process by which nitrogen (N_2) in the atmosphere is converted into ammonia (NH_3). Atmospheric nitrogen or molecular nitrogen (N_2) is relatively inert: it does not easily react with other chemicals to form new compounds. The fixation process frees up the nitrogen atoms from their diatomic form (N_2) to be used in other ways.

Nitrogen fixation, natural and synthetic, is essential for all forms of life because nitrogen is required to biosynthesize basic building blocks of plants, animals and other life forms, e.g., nucleotides for DNA and RNA and amino acids for proteins. Therefore nitrogen fixation is essential for agriculture and the manufacture of fertilizer. It is also an important process in the manufacture of explosives (e.g. gunpowder, dynamite, TNT, etc.). Nitrogen fixation occurs naturally in the air by means of lightning.

Nitrogen fixation also refers to other biological conversions of nitrogen, such as its conversion to nitrogen dioxide. Microorganisms that can fix nitrogen are prokaryotes (both bacteria and archaea, distributed throughout their respective kingdoms) called diazotrophs. Some higher plants, and some animals (termites), have formed associations (symbioses) with diazotrophs.

Biological Nitrogen Fixation

Biological nitrogen fixation was discovered by the German agronomist Hermann Hellriegel and Dutch microbiologist Martinus Beijerinck. Biological nitrogen fixation (BNF) occurs when atmospheric nitrogen is converted to ammonia by an enzyme called nitrogenase. The reaction for BNF is:

$$N_2 + 8\ H^+ + 8\ e^- \rightarrow 2\ NH_3 + H_2$$

The process is coupled to the hydrolysis of 16 equivalents of ATP and is accompanied by the co-formation of one molecule of H_2. In free-living diazotrophs, the nitrogenase-generated ammonium is assimilated into glutamate through the glutamine synthetase/glutamate synthase pathway.

Enzymes responsible for nitrogenase action are very susceptible to destruction by oxygen. Many bacteria cease production of the enzyme in the presence of oxygen. Many nitrogen-fixing organisms exist only in anaerobic conditions, respiring to draw down oxygen levels, or binding the oxygen with a protein such as leghemoglobin.

Microorganisms that Fix Nitrogen (Diazotrophs)

- Cyanobacteria, *e.g.* the highly significant *Trichodesmium*
- Green sulphur bacteria
- *Azotobacteraceae*
- *Rhizobia*
- *Frankia*

Nitrogen Fixation by Cyanobacteria

Cyanobacteria inhabit nearly all illuminated environments on Earth and play key roles in the carbon and nitrogen cycle of the biosphere. In general, cyanobacteria are able to utilize a variety of inorganic and organic sources of combined nitrogen, like nitrate, nitrite, ammonium, urea, or some amino acids. Several cyanobacterial strains are also capable of diazotrophic growth, an ability that may have been present in their last common ancestor in the Archaean. Nitrogen fixation by cyanobacteria in coral reefs can fix twice the amount of nitrogen than on land—around 1.8 kg of nitrogen is fixed per hectare per day. The colonial marine cyanobacterium *Trichodesmium* is thought to fix nitrogen on such a scale that it accounts for almost half of the nitrogen-fixation in marine systems on a global scale.

Root Nodule Symbioses

Legume Family: Plants that contribute to nitrogen fixation include the legume family – Fabaceae – with taxa such as kudzu, clovers, soybeans, alfalfa, lupines, peanuts, and rooibos. They contain symbiotic bacteria called *Rhizobia* within nodules in their root systems, producing nitrogen compounds that help the plant to grow and compete with other plants. When the plant dies, the fixed nitrogen is released, making it available to other plants and this helps to fertilize the soil. The great majority of legumes have this association, but a few genera (e.g., *Styphnolobium*) do not. In many traditional and organic farming practices, fields are rotated through various types of crops, which usually includes one consisting mainly or entirely of clover or buckwheat (non-legume family *Polygonaceae*), which are often referred to as "green manure".

Inga alley farming relies on the leguminous genus Inga, a small tropical, tough-leaved, nitrogen-fixing tree.

Industrial Nitrogen Fixation

The possibility that atmospheric nitrogen reacts with certain chemicals was first observed by Desfosses in 1828. He observed that mixtures of alkali metal oxides and carbon react at high temperatures with nitrogen. With the use of barium carbonate as starting material the first commercially used process became available in the 1860s developed by Margueritte and Sourdeval. The resulting barium cyanide could be reacted with steam yielding ammonia. In 1898 Adolph Frank and Nikodem Caro decoupled the process and first produced calcium carbide and in a subsequent step reacted it with nitrogen to calcium cyanamide. The Ostwald process for the production of nitric acid was discovered in 1902. Frank-Caro process and Ostwald process dominated the industrial fixation of nitrogen until the discovery of the Haber process in 1909.

Haber Process

Artificial fertilizer production is now the largest source of human-produced fixed nitrogen in the Earth's ecosystem. Ammonia is a required precursor to fertilizers, explosives, and other products. The most common method is the Haber process. The Haber process requires high pressures (around 200 atm) and high temperatures (at least 400 °C), routine conditions for industrial catalysis. This highly efficient process uses natural gas as a hydrogen source and air as a nitrogen source.

Much research has been conducted on the discovery of catalysts for nitrogen fixation, often with the goal of reducing the energy required for this conversion. However, such research has thus far failed to even approach the efficiency and ease of the Haber process. Many compounds react with atmospheric nitrogen to give dinitrogen complexes. The first dinitrogen complex to be reported was based on ruthenium,$[Ru(NH_3)_5(N_2)]^{2+}$.

Ambient Nitrogen Reduction

Catalytic chemical nitrogen fixation at temperatures considerably lower than the Haber process is an ongoing scientific endeavour. Nitrogen was converted to ammonia and hydrazine by Alexander E. Shilov in 1970.

Few compounds will cleave the N_2 molecule. Under an atmosphere of nitrogen, lithium metal converts to lithium nitride. Treatment of the resulting nitride gives ammonia. Another example of homolytic cleavage of dinitrogen under mild conditions was published in 1995.

Two equivalents of a molybdenum complex reacted with one equivalent of dinitrogen, creating a triple bonded MoN complex. Since then, this triple bonded complex has been used to make nitriles.

Trimethylsilyl chloride, lithium, and nitrogen molecule react to give tris(trimethylsilyl)amine, under catalysis by nichrome wire or chromium trichloride in tetrahydrofuran.

$$3\ Me_3SiCl + 3\ Li + 1/2\ N2 \rightarrow (Me_3Si)_3N + 3\ LiCl$$

Tris(trimethylsilyl)amine can then be used for reaction with α,δ,ω-triketones to give tricyclic pyrroles.

Catalytic systems for converting nitrogen to ammonia have been developed since the 1980s. In 2003 another was reported based on molybdenum compound, a proton source, and a strong reducing agent. However, this catalytic reduction fixates only a few nitrogen molecules.

Conversion of N_2

The conversion of nitrogen (N_2) from the atmosphere into a form readily available to plants and hence to animals is an important step in the nitrogen cycle, which distributes the supply of this essential nutrient. There are four ways to convert N_2 (atmospheric nitrogen gas) into more chemically reactive forms:

1. Biological fixation: some symbiotic bacteria (most often associated with leguminous plants) and some free-living bacteria are able to fix nitrogen as organic nitrogen. An example of mutualistic nitrogen fixing bacteria are the *Rhizobium* bacteria, which live in legume root nodules. These species are diazotrophs. An example of the free-living bacteria is *Azotobacter*.
2. Industrial N-fixation: Under great pressure, at a temperature of 600 C, and with the use of an iron catalyst, hydrogen (usually derived from natural gas or petroleum) and atmospheric nitrogen can be combined to form ammonia (NH_3) in the Haber-Bosch process which is used to make fertilizer and explosives.
3. Combustion of fossil fuels: automobile engines and thermal power plants, which release various nitrogen oxides (NO_x).
4. Other processes: In addition, the formation of NO from N_2 and O_2 due to photons and especially lightning, can fix nitrogen.

Assimilation

Plants take nitrogen from the soil, by absorption through their roots in the form of either nitrate ions or ammonium ions. All nitrogen obtained by animals can be traced back to the eating of plants at some stage of the food chain.

Plants can absorb nitrate or ammonium ions from the soil via their root hairs. If nitrate is absorbed, it is first reduced to nitrite ions and then ammonium ions for incorporation into amino acids, nucleic acids, and chlorophyll. In plants that have a symbiotic relationship with rhizobia, some nitrogen is assimilated in the form of ammonium ions directly from the nodules. It is now known that there is a more complex cycling of amino acids between *Rhizobia* bacteroids and plants. The plant provides amino acids to the bacteroids so ammonia assimilation is not required and the bacteroids pass amino acids (with the newly fixed nitrogen) back to the plant, thus forming an interdependent relationship. While many animals, fungi, and other heterotrophic organisms obtain nitrogen by ingestion of amino acids, nucleotides and other small organic molecules, other heterotrophs (including many bacteria) are able to utilize inorganic compounds, such as ammonium as sole N sources. Utilization of various N sources is carefully regulated in all organisms.

Ammonification

When a plant or animal dies, or an animal expels waste, the initial form of nitrogen is organic. Bacteria, or fungi in some cases, convert the organic nitrogen within the remains back into ammonium (NH_4^+), a process called ammonification or mineralization. Enzymes Involved are:

- GS: Gln Synthetase (Cytosolic & PLastid)
- GOGAT: Glu 2-oxoglutarate aminotransferase (Ferredoxin & NADH dependent)
- GDH: Glu Dehydrogenase:
  - o Minor Role in ammonium assimilation.
  - o Important in nucleic acid metabolism.

Nitrification

The conversion of ammonia to nitrate is performed primarily by soil-living bacteria and other nitrifying bacteria. In the primary stage of nitrification, the oxidation of ammonium (NH_4^+) is performed by bacteria such as the *Nitrosomonas* species, which converts ammonia to nitrites (NO_2^-). Other bacterial species, such as the *Nitrobacter*, are responsible for the oxidation of the nitrites into nitrates (NO_3^-). It is important for the ammonia to be converted to nitrates because accumulated nitrites are toxic to plant life.

Due to their very high solubility, and because soils are largely unable to retain anions, nitrates can enter groundwater. Elevated nitrate in groundwater is a concern for drinking water use because nitrate can interfere with blood-oxygen levels in infants and cause methemoglobinemia or blue-baby syndrome. Where groundwater recharges stream flow, nitrate-enriched groundwater can contribute to eutrophication, a process that leads to high algal, especially blue-green algal populations. While not directly toxic to fish life, like ammonia, nitrate can have indirect effects on fish if it contributes to this eutrophication. Nitrogen has contributed to severe eutrophication problems in some water bodies. Since 2006, the application of nitrogen fertilizer has been increasingly controlled in Britain and the United States. This is occurring along the same lines as control of phosphorus fertilizer, restriction of which is normally considered essential to the recovery of eutrophied waterbodies.

Denitrification

Denitrification is a microbially facilitated process of nitrate reduction that may ultimately produce molecular nitrogen (N_2) through a series of intermediate gaseous nitrogen oxide products.

This respiratory process reduces oxidized forms of nitrogen in response to the oxidation of an electron donor such as organic matter. The preferred nitrogen electron acceptors in order of most to least thermodynamically favourable include nitrate (NO_3^-), nitrite (NO_2^-), nitric oxide (NO), nitrous oxide (N_2O) finally resulting in the production of dinitrogen (N_2) completing the nitrogen cycle.

The process is performed primarily by heterotrophic bacteria (such as *Paracoccus denitrificans* and various pseudomonads), although autotrophic denitrifiers have also been identified (e.g., *Thiobacillus denitrificans*). Denitrifiers are represented in all main phylogenetic groups. Generally several species of bacteria are involved in the complete reduction of nitrate to molecular nitrogen, and more than one enzymatic pathway have been identified in the reduction process.

Direct reduction from nitrate to ammonium, a process known as dissimilatory nitrate reduction to ammonium or DNRA, is also possible for organisms that have the nrf-gene. This is less common than denitrification in most ecosystems as a means of nitrate reduction. Other genes known in microorganisms which denitrify include *nir* (nitrite reductase) and *nos* (nitrous oxide reductase) among others; organisms identified as having these genes include *Alcaligenes faecalis*,

Alcaligenes xylosoxidans, many in the *Pseudomonas* genus, *Bradyrhizobium japonicum*, and *Blastobacter denitrificans*.

Nutrient Limitation

All organisms require certain nutrients in their surroundings (available to them) for survival. Depending upon the ecosystem, nitrogen is most likely the limiting nutrient, although phosphorus is the other primary limiting nutrient and these two elements interact chemically. Some organisms appear to be able to denitrify and remove phosphorus. The triple bond of N_2 makes this a very stable compound; most organisms (i.e. plants) depend upon others to break this down to make it available for biochemical reactions. Symbiotic relationships between *Rhizobium* species and legumes are well-documented.

Conditions Required

Denitrification takes place under special conditions in both terrestrial and marine ecosystems. In general, it occurs where oxygen, a more energetically favourable electron acceptor, is depleted, and bacteria respire nitrate as a substitute terminal electron acceptor. Due to the high concentration of oxygen in our atmosphere denitrification only takes place in anoxic environments where oxygen consumption exceeds the oxygen supply and where sufficient quantities of nitrate are present. These environments may include certain soils and groundwater, wetlands, oil reservoirs, poorly ventilated corners of the ocean, and in seafloor sediments.

Denitrification generally proceeds through some combination of the following intermediate forms:

$$NO_3^- \rightarrow NO_2^- \rightarrow NO + N_2O \rightarrow N_2\ (g)$$

The complete denitrification process can be expressed as a redox reaction:

$$2\ NO_3^- + 10\ e^- + 12\ H^+ \rightarrow N_2 + 6\ H_2O$$

This reaction shows a fractionation in isotope composition. Lighter isotopes of nitrogen are preferred in the reaction, leaving the heavier nitrogen isotopes in the residual matter. The process can cause delta-values of up to "40, where delta is a representation of the difference in isotopic composition. This can be used to identify denitrification processes in nature.

Deliberate Use of Process

Denitrification is commonly used to remove nitrogen from sewage and municipal wastewater. It is also an instrumental process in wetlands and riparian zones for the removal of excess nitrate from groundwater resulting from excessive agricultural or residential fertilizer usage.

Reduction under anoxic conditions can also occur through process called anaerobic ammonium oxidation (anammox):

$$NH_4^+ + NO_2^- \rightarrow N_2 + 2\ H_2O$$

In some wastewater treatment plants, small amounts of methanol, ethanol, acetate, glycerin, or proprietary products are added to the wastewater to provide a carbon source for the denitrification bacteria. Denitrification processes are also used in the treatment of industrial wastes.

Influence on Global Climate Change

Increasing carbon dioxide levels within the atmosphere will influence global nutrient cycling, yet it is difficult to predict what those interactions might be. Chemical interactions between soils and the atmosphere will be influenced by changes in atmospheric composition. There are indications that increased fertilization of soils with nitrogen causes a decrease in carbon sequestration.

It has been shown that streams and rivers receiving nitrogen inputs from urban and agricultural land uses are a significant source of nitrous oxide to the atmosphere.

4

Plant-Soil Interactions: Nutrient Uptake

Plant growth and development largely depend on the combination and concentration of mineral nutrients available in the soil. Plants often face significant challenges in obtaining an adequate supply of these nutrients to meet the demands of basic cellular processes due to their relative immobility. A deficiency of any one of them may result in decreased plant productivity and/or fertility. Symptoms of nutrient deficiency may include stunted growth, death of plant tissue, or yellowing of the leaves caused by a reduced production of chlorophyll, a pigment needed for photosynthesis. Nutrient deficiency can have a significant impact on agriculture, resulting in reduced crop yield or reduced plant quality. Nutrient deficiency can also lead to reduced overall biodiversity since plants serve as the producers that support most food webs.

Changes in the climate and atmosphere can have serious effects on plants, including changes in the availability of certain nutrients. In a world of continual global climate change, it is important to understand the strategies that plants have evolved to allow them to cope with some of these obstacles.

Two classes of nutrients are considered essential for plants: macronutrients and micronutrients. Macronutrients are the building blocks of crucial cellular components like proteins and nucleic acids; as the name suggests, they are required in large quantities. Nitrogen, phosphorus, magnesium, and potassium are some of the most important macronutrients. Carbon, hydrogen, and oxygen are also considered macronutrients as they are required in large quantities to build the larger organic molecules of the cell; however, they represent the non-mineral class of macronutrients. Micronutrients, including iron, zinc, manganese, and copper, are required in very small amounts. Micronutrients are often required as cofactors for enzyme activity.

Mineral nutrients are usually obtained from the soil through plant roots, but many factors can affect the efficiency of nutrient acquisition. First, the chemistry and composition of certain soils can make it harder for plants to absorb nutrients. The nutrients may not be available in certain soils, or may be present in forms that the plants cannot use. Soil properties like water content, pH, and compaction may exacerbate these problems.

Second, some plants possess mechanisms or structural features that provide advantages when growing in certain types of nutrient limited soils. In fact, most plants have evolved nutrient uptake mechanisms that are adapted to their native soils and are initiated in an attempt to overcome nutrient limitations. One of the most universal adaptations to nutrient-limited soils is a change in root structure that may increase the overall surface area of the root to increase nutrient acquisition or may increase elongation of the root system to access new nutrient sources. These changes can lead to an increase in the allocation of resources to overall root growth, thus resulting in greater root to shoot ratios in nutrient-limited plants (Lopez-Bucio *et al.*, 2003).

Plants are known to show different responses to different specific nutrient deficiencies and the responses can vary between species. The most common changes are inhibition of primary root growth (often associated with P deficiency), increase in lateral root growth and density (often associated with N, P, Fe, and S deficiency) and increase in root hair growth and density (often associated with P and Fe deficiency).

Plant roots exhibit a variety of changes in response to nutrient deficiency, including inhibition of primary root elongation and increased growth and density of lateral roots and root hairs. These responses are species-, genotype-, and nutrient-specific, but they are generalized in this figure to demonstrate all potential effects.

While nutrient deficiencies can pose serious threats to plant productivity, nutrients can become toxic in excess, which is also problematic. When some micronutrients accumulate to very high levels in plants, they contribute to the generation of reactive oxygen species (ROS), which can cause extensive cellular damage. Some highly toxic elements like lead and cadmium cannot be distinguished from essential nutrients by the nutrient uptake systems in the plant root, which means that in contaminated soils, toxic elements may enter the food web via these nutrient uptake systems, causing reduced uptake of the essential nutrient and significantly reduced plant growth and quality.

In order to maintain nutrient homeostasis, plants must regulate nutrient uptake and must respond to changes in the soil as well as within the plant. Thus, plant species utilize various strategies for mobilization and uptake of nutrients as well as chelation, transport between the various cells and organs of the plant and storage to achieve whole-plant nutrient homeostasis. Here, we briefly describe a few examples of strategies used by plants to acquire nutrients from the soil.

Plant Acquisition of Nutrients: Direct Uptake from the Soil

Potassium. Potassium (K) is considered a macronutrient for plants and is the most abundant cation within plant cells. Potassium has a number of important functions within plants, including balancing the charges of cellular anions, enzyme activation, control of stomatal opening/closing and serving as an osmoticum for cellular growth.

Potassium deficiency occurs frequently in plants grown on sandy soils resulting in a number of symptoms including browning of leaves, curling of leaf tips and yellowing (chlorosis) of leaves, as well as reduced growth and fertility.

Potassium uptake processes have been the subject of intense study for several decades. Early studies indicated that plants utilize both high and low affinity transport systems to directly acquire potassium from the soil. Low affinity transport systems generally function when potassium levels in the soil are adequate for plant growth and development. This process is mediated by ion channels in the plasma membrane of root cells, allowing passive transport of K^+ from areas of relatively high external concentration into the plant cells where the concentration of K^+ is lower. The expression of these low affinity transporters does not appear to be significantly affected by potassium availability.

Under conditions of potassium limitation, in contrast, plants usually induce high affinity K^+ transport systems. There are likely many proteins involved in high affinity potassium transport, but in Arabidopsis, two proteins have been identified as the most important transporters in this process. Interestingly, one of these transporters, AtHAK5, is a carrier protein and thought to mediate active transport of potassium into plant roots, whereas the other protein, AKT1, is a channel protein and likely mediates a passive transport mechanism with an increased affinity for K^+ under conditions of potassium limitation (Pyo *et al.*, 2010). More recent work shows that plants contain a number of different transport systems to acquire potassium from the soil and distribute it within the plants. Although much remains to be learned about potassium uptake and translocation in plants, it is clear that the mechanisms involved are complex and tightly controlled to allow the plant to acquire sufficient amounts of potassium from the soil under varying conditions.

Iron. Iron is essential for plant growth and development and is required as a cofactor for proteins that are involved in a number of important metabolic processes including photosynthesis and respiration. Despite the fact that iron is the fourth most abundant element in the earth's crust, it is often limiting for plants due to the fact that it tends to form insoluble complexes in aerobic soils of neutral to basic pH (Guerinot & Yi, 1994). It is thought that iron limitation is a problem for plants on as much as 30% of soils worldwide. Iron-deficient plants often display interveinal chlorosis, in which the veins of the leaf remain green while the areas between the veins are yellow. Due to the limited solubility of iron in many soils, plants often must first mobilize iron in the rhizosphere (a region of the soil that surrounds, and is influenced by, the roots) before transporting it into the plant. Two distinct mechanisms have evolved that are utilized by plants to acquire iron from the soil, termed the Strategy I and Strategy II responses (Connolly & Walker, 2008).

Strategy I is used by all plants except the grasses. It is characterized by three major enzymatic activities that are induced in response to iron limitation and that are located at the plasma membrane of cells in the outer layer of the root. First, strategy I plants induce the activity of H^+-ATPase, which uses the energy of ATP to pump protons out of the root cells and into the rhizosphere. This activity thus serves to acidify the rhizosphere and the drop in rhizosphere pH solubilizes ferric iron (Fe^{3+}), making it more available for subsequent uptake by the plant. Second, strategy I plants induce the activity of a plasma-membrane-bound ferric chelate reductase. Ferric chelate reductase activity reduces ferric iron to the more soluble ferrous form (Fe^{2+}) of iron. Finally, plants induce the activity of a ferrous iron transporter that moves ferrous iron across the plasma membrane and into the plant.

In contrast, the grasses utilize strategy II to acquire iron under conditions of iron limitation. Following the imposition of iron limitation, strategy II species begin to synthesize special molecules called phytosiderophores (PSs) that display high affinity for ferric iron. PSs are secreted into the rhizosphere where they bind tightly to ferric iron. Finally, the PS-ferric iron complexes are transported into root cells by PS-Fe(III) transporters. Interestingly, while both strategies are relatively effective at allowing plants to acquire iron from the soil, the strategy II response is thought to be more efficient because grass species tend to grow better in calcareous soils (which have a high pH and thus have limited iron available for uptake by plants).

A. Strategy I plants induce the activity of a proton ATPase, a ferric chelate reductase, and a ferrous iron transporter when faced with iron limitation. B. In contrast, Strategy II plants synthesize and secrete phytosiderophores (PS) into the soil in response to iron deficiency. The PSs bind ferric iron with high affinity and the resulting PS-$Fe3^{+}$ complexes are transported from the soil into the plant root.

Plant Acquisition of Nutrients: Symbioses with Soil-based Microorganisms

Nitrogen and phosphorus are among the elements considered most limiting to plant growth and productivity because they are often present in small quantities locally or are present in a form that cannot be used by the plant. As a result, the evolution of many plant species has included the development of mutually beneficial symbiotic relationships with soil-borne microorganisms. In these relationships, both the host plant and the microorganism symbiont derive valuable resources that they need for their own productivity and survival as a result of the association.

Nitrogen Fixation. Despite the fact that nitrogen is the most abundant gaseous element in the atmosphere, plants are unable to utilize the element in this form (N_2) and may experience nitrogen deficiency in some soils that have low nitrogen content. Since nitrogen is a primary component of both proteins and nucleic acids, nitrogen deficiency imposes significant limitations to plant productivity. In an agricultural setting, nitrogen deficiency can be combated by the addition of nitrogen-rich fertilizers to increase the availability of nutrients and thereby increase crop yield. However, this can be a dangerous practice since excess nutrients generally end up in ground water, leading to eutrophication and subsequent oxygen deprivation of connected aquatic ecosystems.

Plants are able to directly acquire nitrate and ammonium from the soil. However, when these nitrogen sources are not available, certain species of plants from the family Fabaceae (legumes) initiate symbiotic relationships with a group of nitrogen fixing bacteria called Rhizobia. These interactions are relatively specific and require that the host plant and the microbe recognize each other using chemical signals. The interaction begins when the plant releases compounds called flavanoids into the soil that attract the bacteria to the root. In response, the bacteria release compounds called Nod Factors (NF) that cause local changes in the structure of the root and root hairs. Specifically, the root hair curls sharply to envelop the bacteria in a small pocket. The plant cell wall is broken down and the plant cell membrane invaginates and forms a tunnel called an infection thread that grows to the cells of the root cortex. The bacteria become wrapped in a plant derived membrane as they differentiate into structures called bacteroids. These structures are allowed to enter the cytoplasm of cortical cells where they convert atmospheric nitrogen to ammonia, a form that can be used by the plants. In return, the bacteroids receive photosynthetically derived carbohydrates to use for energy production (reviewed by Limpens & Bisseling, 2003; Ferguson *et al.* 2010).

A. Process of root cell colonization by rhizobacteria. B. Nodule formed by nitrogen fixing bacteria on a root of a pea plant (genus Pisum).

Mycorrhizal Interactions with Plants

In addition to symbiotic relationships with bacteria, plants can participate in symbiotic associations with fungal organisms as well. Current estimates of the frequency of plant-mycorrhizal associations suggest that around 80% of all plants establish some type of mycorrhizal symbiosis, and many studies indicate that these relationships are millions of years old (Karandashov & Bucher, 2005; Vance, 2001). There are several classes of mycorrhiza, differing in structural morphology, the method of colonizing plant tissue, and the host plants colonized. However, there are two main classes that are generally regarded as the most common and therefore, the most ecologically significant. The endomycorrhizae are those fungi that establish associations with host plants by penetrating the cell wall of cortical cells in the plant roots. By contrast, ectomycorrizae develop a vast hyphae network between cortical cells but do not actually penetrate the cells.

The most common endomycorrhizal interaction occurs between arbuscular mycorrhizal fungi (AMF; also called Vesicular-Arbuscular

Mycorrhiza or VAM) and a variety of species of grasses, herbs, trees and shrubs. When phosphate is available in the soil, plants are able to acquire it directly via root phosphate transporters. However, under low phosphate conditions, plants become reliant on interactions with mycorrhizal fungi for phosphorus acquisition. Mycorrhizal spores present in the soil are germinated by compounds released from the plant. Hyphae extend from the germinating spore and penetrate the epidermis of the plant root. Inside the root, the hyphae branch and penetrate cortical cells, where highly branched structures called arbuscules develop. Externally, hyphae extend into the soil beyond the area accessible to the root. This kind of symbiosis facilitates plant phosphorus uptake from the soil by increasing the root's absorptive surface area. Since plants take up phosphorus at a much higher rate than phosphorus diffuses into the soil surrounding the root, a phosphorus depletion zone is quickly established, limiting uptake of phosphorus by the plant. However, AMF hyphae form a bridge between the internal root environment and the area beyond this depletion zone enabling the plant to acquire significantly more phosphorus through its symbiotic partner than it could on its own (reviewed by Karandashov & Bucher 2005).

A. Diagram of arbuscular mycorrhizae colonization of a plant root showing the extension of hyphae beyond the phosphorus depletion zone and the presence of arbuscules in cells of the root cortex. B. Diagram of Ectomycorrhizal fungi showing growth of hyphae around cortical cells, a mantle sheath on the outside of the root, and hyphae that extend into soil around the root.

Ectomycorrhizal fungi (EcM) form associations with many tree species (birch, oak, spruce, pine, fir), making these microorganisms an important component of the biodiversity found in forest ecosystems, particularly temperate and boreal forests in the northern hemisphere. Although the molecular mechanisms underlying initiation of EcM-plant interactions are not well defined, it is clear that signalling molecules are released by both symbiotic partners, resulting in fungal hyphae growth towards compatible plant roots. The EcM forms an extensive hyphal sheath, or mantle, around the root cap and a hartig net of hyphae that extends into the root itself and surrounds the cells within the root cortex. Hyphae from the mantle extend into the surrounding soil, providing availability and storage of nutrients like phosphorus and nitrogen for the plant. This is particularly important for trees growing in forest soils because a large portion of the nutrients available are located in the litter layer and are not accessible to the roots of these trees. The EcM produce enzymes that can digest this organic material and also have the ability to mobilize the nutrients to the hartig net, making them available to the plant. In this way, the interaction between trees and EcM is not only beneficial, but in many cases, is essential (reviewed by Nehls *et al.*, 2008).

As is the case with bacterial symbioses, the benefit to the fungal partners in these relationships involves transfer of photosynthetic carbohydrates from the plant to the microorganism for metabolism and energy production. However, the nutritional advantage for plants of such symbiotic relationships far outweighs the energetic cost imposed by hosting microorganisms like rhizobacteria, AMF, and EcM.

Summary

Although plants are non-motile and often face nutrient shortages in their environment, they utilize a plethora of sophisticated mechanisms in an attempt to acquire sufficient amounts of the macro- and micronutrients required for proper growth, development and reproduction. These mechanisms include changes in the developmental programme and root structure to better "mine" the soil for limiting nutrients, induction of high affinity transport systems and the establishment of symbioses and associations that facilitate nutrient uptake. Together, these mechanisms allow plants to maximize their nutrient acquisition abilities while protecting against the accumulation of excess nutrients, which can be toxic to the plant. It is clear that the ability of plants to utilize such mechanisms exerts significant influence over crop yields as well as plant community structure, soil ecology, ecosystem health, and biodiversity.

How Do Plants Absorb Mineral Nutrients?

Soils usually contain the essential mineral elements at concentrations much lower than is required in plant tissues. In addition, some soils contain high concentrations of mineral elements not needed by plants. The mechanisms by which plant roots take up nutrients must therefore be selective and enable the accumulation of nutrients against a steep concentration gradient. Selectivity of uptake enables plants to regulate the composition of elements in their tissues.

Roots - the Primary Site of Nutrient Aquisition by Plants

Plant roots can penetrate to significant depths and are branched so that they can explore large volumes of soil. Some nutrients (eg. Nitrate) are only "weakly bound" in the soil, and can flow towards the root surface when a plant uses water. Other nutrients (eg.

Phosphate) can be tightly bound in some soils, thus, continued uptake by plants relies on growth of roots into new areas of the soil. Root hairs, which greatly increase the volume of soil explored adjacent to the main root and also increase the root's surface area, may be particularly active in nutrient absorption. The degree of phosphate adsorption by soil particles depends on the soil type; sands adsorb phosphate only weakly whereas clays rich in iron oxides absorb it very strongly.

Generally, grasses have extensive root systems with many fine branches and sub-branches. These features contribute to good soil exploration for nutrients. Other plants, such as several legume species, have a thick tap-root and relatively fewer fine roots. Legumes form symbiotic relationships (both partners benefit) with selected species of soil bacteria and fungi. The bacteria (Rhizobia) form root nodules and fix nitrogen gas from the atmosphere and provide it in a form suitable for plants. Mycorrhizal fungi form a network of hyphae extending into the soil to facilitate phosphorus uptake. In return the host plant supplies the nodule bacteria and mycorrhizal fungi with sugars fixed via photosynthesis.

How is Selectivity of Nutrient Uptake Achieved?

Mineral nutrients are absorbed by plants from the soil solution as ions (eg. K+, NO3-, HPO42-, SO42-). An ion is the charged particle formed by the removal or addition of electrons to any particular atom. The ions have two possible pathways for roots uptake: (i) through the apoplasm (cell walls and intercellular spaces) and/or (ii) movement cell-to-cell in the symplasm (through the living cells). The apoplasmic pathway is blocked at the endodermis by "casparian bands" (deposits of hydrophobic materials) in the cell walls. These apoplasmic barriers force the water and ions to move through cellular membranes if they are to be absorbed by the root.

The "casparian bands" that block apoplasmic movement of ions are essential to selectivity since this is achieved by specific ion transporter proteins embedded in the membranes of plant cells. Membranes are composed of a lipid bilayer which prevents the free movement of substances in and out of cells. The transporter proteins are highly specific for the transfer of different ions across the membrane. So, for any given nutrient ion to be accumulated by a plant it must be transported across root cell membranes. Regulation of these transporters enables control of internal nutrient composition (ie. selectivity) and concentrations.

Ions absorbed by the roots are then loaded into the xylem (conducting vessels in the roots, stems, and leaves) and transported with water up to the leaves when the plant transpires.

Factors Affecting Nutrient Uptake

Several soil and plant factors influence the rates of nutrient uptake by roots. Soil factors such as low water availability, low nutrient availability, low temperature, and low (acidic) or high (alkaline) pH can all decrease rates of nutrient uptake. Soil pH is an important parameter since pH affects the solubility and ionic form of particular nutrients. In addition, any soil constraints to root growth (eg. compaction, disease, waterlogging, etc.) can also diminish rates of nutrient uptake by plants. Many of the soil factors listed above can, to some degree, be controlled by the turf manager.

Plant factors, as determined by genotype, that influence nutrient uptake rates include: rooting patterns, ability to form symbiotic relationships, rhizosphere chemistry, and sensitivity of roots to several environmental constraints. Selection of the most appropriate genotypes, especially for tolerance to soil constraints such as high or low pH, can benefit turf culture and nutrient management on soils with these problems.

Optimizing Nitrogen Use on the Farm

When aiming to optimize yield and protein of wheat grown under rainfed conditions, growers must supply sufficient, not excessive, nitrogen. What works one year may not necessarily be right the next year. The variability of rainfall amount and distribution results in highly variable responses to a set rate of N-fertilizer.

Growers should aim to match the supply of nitrogen, mineralized from soil organic matter or from a bag of fertilizer, with the requirements of the crop. Crop requirements constantly change depending on available soil water and rainfall.

Which Farms Could Benefit from These Nitrogen Trials?

- Those where little or no nitrogen is applied to crops either directly or via rotation crops;
- those farms with lower yields in most years than surrounding farms;
- those farms with significantly lower yields than reported from regional research trials;

- those where symptoms of haying-off appear;
- those with variable rainfall during the season and across years;
- those farms that consistently produce grain with protein contents lower than required.

Too Little Nitrogen and Low Yield

Many farms in the Mediterranean region have low wheat yields of around 1 t/ha. For a high proportion of these farms yield could be substantially boosted by the application of nitrogen either from a sack or through use of legume crops. Nitrogen is not commonly applied at all or applied spasmodically at one or two sacks per hectare. Fertility is seemingly maintained by the use of fallows. It is on these farms that basic nitrogen trials paralleled by economic assessments (cost-benefit or gross margins analyses) are most needed.

Too Much Nitrogen and Haying-off

Farmers growing cereals where rainfall is both limited and variable in distribution must also be very careful not to apply too much nitrogen. When water is adequate, nitrogen stimulates the crop to grow faster and accumulate more biomass and set a potentially high yield. If water then becomes inadequate to support that increased biomass as the season progresses, the crop responds by shedding leaf and other tissues solely to survive. This is called 'haying-off'.

Van Herwaarden

Depending on its severity it can cause serious economic loss to growers on three fronts; it decreases yield, the harvested crop is poor quality because of pinched or shrivelled grain and the applied nitrogen was wasted. Indeed, the costly applied nitrogen was more than wasted because it was the cause of the yield losses. At the time of flowering, crops of very high-N status contain fewer reserves of water-soluble carbohydrates (stem sugars) than crops of low-N status. Shortage of these stem sugars, normally transferred to the ear to fill the grain, is responsible for the yield reduction and pinched grain of very high N crops.

Very high levels of nitrogen can over-stimulate tillering which locks up the carbohydrates in structural materials rather than leaving them in storage to be used later to fill the grains.

Balancing Nitrogen to Water Availability

In those seasons when there is sufficient rain during spring to ensure there is little or no drought stress during the later stages of stem elongation through to grain filling, a higher nitrogen crop has a higher yield potential.

It can retain more green leaves for longer and use them to fill the grain from current photosynthesis. It only needs to call on stored stem sugars during short periods of very high requirement.

By contrast, in the event of post-flowering drought, these crops of high-N status suffer the combined stresses of the drought (haying-off) and shortage of stem sugar reserves. Crops of low nitrogen status grow less before flowering thus having relatively fewer grains to fill per unit land area and less above ground biomass to supply with water after flowering.

Consequently, during a late drought they lose relatively little of their green leaf. This green leaf maintains current photosynthesis and this assimilate, when combined with the high reserves of stem sugars, fills the grains satisfactorily through the drought. If the drought is particularly severe causing green leaves to die, the stem reserves alone can go a long way towards filling the grains.

When there is adequate water throughout grain filling, applied nitrogen boosts yield. However, when a drought occurs, then extra N may diminish yield. How should farmers decide what level of nitrogen they should add to boost yield in good years but not cause haying-off in bad years?

The solution is a series of collaborative on-farm trials with inputs measured and matched by calculation to requirements. Trials should span a region and be carried out preferably over two or more years. Spanning a region through time enables questions to be answered about the impact of soil type on nitrogen requirements and conjointly the impact of different environmental factors.

Crop Management to Maximize the Benefit of Supplementary Nitrogen

Wheat can be managed to produce grain at or close to the yield ceiling set by water availability in each season.

It requires knowledge of what factors may be affecting yield other than nitrogen. It requires nitrogen budgeting before sowing and further budgeting as the season progresses. Each decision about whether or not to add N with time depends on cumulative and current rainfall and what targets have been set for yield and protein.

Nitrogen will have little impact on crop yield if other factors present a greater limitation. For a good yield response to nitrogen attend first to the following

Criteria to Satisfy First to Realize Benefits from N Applications

- Sow locally adapted cultivars at the optimum time;
- ensure there is a low risk of disease;
- adjust soil pH if possible to levels where it presents little or no limitation to growth;
- ensure nutrients such as phosphorus, zinc and trace elements are not limiting growth;
- ensure there are no subsoil constraints to root growth such as sodicity or trace element toxicity.

Discussions with Collaborating Farmers

A good place to start your discussions with collaborating farmers is to find out, in general terms, to what extent the above-mentioned conditions have been met over recent years.

As the proposed trials are about nitrogen, ask about direct applications of nitrogen to wheat in the last three seasons.

Even without details you will soon be able to categorize a farmer as a non-user, a light haphazard user, a heavy user, a user who monitors the crop's requirements and matches N applications to them. Your aim is to persuade the associated farmer group into that last category.

Ask about yields and whether farmers are happy with their farms' output. If yields approximate 1 t/ha, the farm is likely to be one of low input and a good candidate for a very basic N study, perhaps broadcasting different levels of N in strips within the farmer's crop (described later). If the farm produces higher yields, you might have to contemplate the more complex N trials also described later.

Expand your discussions to the other criteria above. It is likely that the haphazard N-user will have paid little attention to other nutrients and other constraints. You might need to help with a soil test.

Find out whether farmers use crop sequences that include a legume for harvesting or a legume for ploughing back in to improve soil organic matter and fertility.

You may have discussed the earlier figure of N-response to demonstrate how yield and protein can change with N application, but it might be worth explaining that this is not the only possible response. The shape and level of the curves is dependent on many other interacting factors that may or may not be present on the farm.

You might like to use these figures to point out how important it is to do a basic low-effort study to categorize the likely response to N on that farm as no exact pattern can be assumed.

Note also the different responses in protein because an input for that will be required when you do your N budgeting.

Nitrogen Budgeting for a Farm

When you have made your assessment of the nutrition state of the farm and have some idea of the likely main constraints to yield, the next step is to construct your nitrogen budget for the farm.

If you can arrange a N soil test before planting and complete the following budget, you will be able to calculate N fertilizer requirements at sowing that will be accurate enough to meet targets of 70-80 percent of average yield assuming 11 percent grain protein.

More accurate updated assessments of nitrogen requirements can be made as the season progresses. Actions to rebalance N applications to crop requirements might, for example, be taken when the stems of the crop begin to elongate. This should be done to increase yield and protein but only as a tactical response to favourable soil water supply and low crop N status. It should not be done as a matter of course. A second tactical application to further increase protein can be made up to flowering, but this will also depend on water supply.

How Much N Does the Crop Require?

Start by calculating overall crop nitrogen requirements assuming a target for yield that is the average for recent years on the farm.

The following example calculation estimates the nitrogen requirement for a wheat crop yielding 2 t ha^{-1} of grain at 11 percent protein. It uses a correction factor of 2.34 that converts percent protein to kilograms of nitrogen and assumes that 25 percent of crop nitrogen is held in the straw.

1) Target crop yield x target protein percent x correction factor = nitrogen required over the season

e.g. 2 t ha^{-1} x 11 percent x correction factor = 52 kg N ha^{-1} required.

Similarly, a 1 t ha^{-1} crop would require 1 t ha^{-1} x 11 percent x 2.34 = 25.7 kg N ha^{-1} from somewhere.

How Much N is Already in the Soil?

The above-mentioned calculation might tell how much nitrogen the crop requires but a part of that may not need to be applied as fertilizer, as it may already be available in the soil. To test for existing soil mineral nitrogen, take several soil cores to 60 cm from around the field. This is the extent to which many roots penetrate. Tests to 10 cm are of limited use but if it has been dry since the last harvest, 30 cm may be deep enough to allow for most mineral N.

Send the samples to a testing agency or use the method of Wetselaar and colleagues for estimating soil nitrate levels in the field. For the example calculation it is assumed that the tests indicated 20 kg N ha^{-1} of mineral nitrogen in the soil to a depth of 60 cm.

Remember that mineralization is a continuing process so will make additional nitrogen available throughout the growing season. Nitrogen mineralization varies between 60 kg ha^{-1} for infertile soils with less than 0.9 percent organic carbon to 100 kg ha^{-1} on fertile soils (more than 1.8 percent organic carbon). The calculation assumes the soil is infertile.

Existing soil mineral N + N mineralized during the season = total season supply in the soil e.g. 20 kg N ha^{-1} + 60 kg N ha^{-1} = 80 kg N ha^{-1} potentially available to crop.

However, actual recovery of nitrogen from soil supplies is often closer to 50 percent than 100 percent e.g. 0.5 x 80 kg N ha^{-1} = 40 kg N ha^{-1} actually available to the crop during the season.

The deficit in soil N available for the crop is (required N - available soil N) or ((1) - (3)). In the example for a crop of 2 t ha^{-1} this is 52 kg N ha^{-1} - 40 kg N ha^{-1} = 12 kg N ha^{-1}.

As only about 50 percent of applied fertilizer is available in the first season, for our example, the estimated requirement for fertilizer to be applied is 12 kg N ha^{-1} x 2 = 24 kg N ha^{-1}.

For a 1 t ha^{-1} crop the entire requirement could be met by soil mineralization. In fact, mineralization alone could support a crop of 1.5 t ha^{-1} providing all other management options were optimized (e.g. correct seedbed preparation leading to good crop establishment).

Setting up a High Yield Potential by Splitting N Applications

Managing nitrogen fertilizer during the crop's growth can be done so that the proportions of the crop yield components are changed.

For example, the later in crop growth that N is applied the greater is its impact on increasing protein and the less its proportional impact on biomass and yield.

Budget to work out how much nitrogen the crop needs and its sources			
		calculation	***example***
Crop N needs			
A	target yield for crop		2 t ha^{-1}
B	target protein %		11 %
C	correction factor of 2.34		
D	crop N needs	(A x B x C)	52 kg N ha^{-1}
Soil N supply			
E	mineral N at sowing to 60 cm		20 kg N ha^{-1}
F	mineralization of soil N during season		60 kg N ha^{-1}
G	gross N supply	(E + F)	80 kg N ha^{-1}
H	net N supply assuming 50% available to crop	(G) x 0.5	40 kg N ha^{-1}
N the farmer should add			
J	extra N needed	(D - H)	12 kg N ha^{-1}
K	only 50% applied to crop so N to add is -	1 x 2	24 kg N ha^{-1}

If the proposed N-budgeting method is used to determine N-needs of a crop of 3 t ha^{-1} at 13.5 percent protein and apply all that nitrogen at sowing, a yield potential of approximately 4 t ha^{-1} at 10 percent protein is actually set up.

This higher yield potential, when set early, has the detrimental effects of increasing crop height, reducing stem strength and using more soil water up to flowering.

This approach can increase the bulk of a crop in wet years to the extent that it lodges and in dry years leads to haying-off. The consequence in both cases is that the potential is lost.

Actually achieving the potential of 3 t ha^{-1} at 13.5 percent protein without lodging or haying-off requires the budgeted amount of nitrogen to be applied through time, with only part applied before or at sowing.

In effect this fools the crop into initially underestimating its potential so it then sets up its first yield components conservatively (i.e. fewer tillers, less leaf area and spikes per unit area).

Later in the season when conditions improve (by the remaining budgeted nitrogen being applied), these yield components are no longer plastic so the crop is restricted to using the later yield components of kernels per spike and kernel weight to express its yield potential.

When kernel weight is the only plastic component left, any late nitrogen available or applied has to be used up by the crop in increasing protein.

Encouraging the crop to arrange its form into the yield components that the farmer wants while ensuring growth is balanced to the environment, is a continuing exercise throughout the season. At the time when a decision needs to be made whether to apply supplementary N-fertilizer, available soil water should be assessed.

As a guide, if rainfall from sowing up to this time exceeds the potential evaporation then the storage of water in the soil is likely to have increased and the chance of getting a positive response to N-fertilizer applied at this time is high.

If rainfall has been low or absent then response will probably be small. It should be borne in mind that nitrogen is used much more efficiently if applied just before a rainfall event.

If management decisions or the environment seriously limited early growth, N applied late may not compensate for low yield potential. An adequate plant stand must first be established to make it worthwhile applying N during later growth.

Interpreting the Effects of N Fertilizer Applied at Sowing: A Case Study

The following trial shows that nitrogen applied at sowing can have positive or negative effects on yield depending on the availability of water at different stages of development. It shows that you have to be very careful with nitrogen. It was carried out in a relatively dry region.

The six wheat cultivars used differed very little in duration, all reaching anthesis over an eight-day period, but this turned out to be very important to yield.

On average across the six cultivars, an application of 37 kg N ha^{-1} before sowing increased yield by about 200 kg ha^{-1}. This ranged from a 350 kg ha^{-1} yield increase for the two earliest flowering cultivars to a yield decrease of 200 kg ha^{-1} for the two latest cultivars.

Why Did N Not Always Increase Yield?

A yield increase would normally be expected in all six cultivars from this small amount of N fertilizer, *but the weather complicated the response.* All crops were exposed to a hot, very dry period from immediately preceding anthesis and into early grain filling so their soil water reserves were significantly depleted. Water, not N, became the major constraint to yield.

The cultivars that flowered first and very early in the drought had, in effect, six more days of water to use on filling grain than the latest flowering wheats. They could fill the extra grain sites that had resulted from the 37 kg N ha^{-1} applied before sowing.

By contrast, the latest two cultivars ran out of water for completing grain filling and ran out earlier in the treatments that had received N at sowing.

They ran out earlier because the added N had stimulated tillering and leaf area production. This in turn had led to faster and greater total use of stored soil moisture before anthesis.

This trial was chosen to demonstrate that there are many complex interactions occurring in crops. In this case, a failure to respond to N was nothing to do with the genotype's inherent responsiveness to N, but due to a third factor that was dominant in that particular season.

Always be alert for such additional variables. Repeating trials over several seasons is often the only way to average out their impact when optimizing a farming package for a location.

Setting up Your on-farm Trials

A nominal N Response Curve

If you are dealing with collaborating farmers who have not applied N previously, or used it sporadically, you need to determine a nominal response curve to N for the location.

Make this a low effort study within the farmers' crops and use areas within different planting dates and different varieties if those are available.

Use a strip design with 0, 25, 50, 75, 100 kg N ha^{-1} applied in bands like stripes on a scarf in each of the areas. Mark out the plots and areas clearly with handfuls of lime. A single replicate using large plots of say 10m x 10m, should be sufficient for this starter trial.

Measure only grain yield at maturity but encourage farmers to measure rainfall throughout the season. This will affect the outcome as in the case study. You will be able to show the costs of the additional nitrogen against the value of the increments in crop yield. On that basis you will together be able to decide whether further, more complex, optimizing studies are worth doing.

Beyond the Nominal Response Curve

For farmers who have used N and have some idea that it may be beneficial to crops, you might like to consider suggesting one of the following three trials or some variant.

The three trials have the same basic design but add increasing levels of complexity.

Complexity increases as more possible limitations to crop growth are considered (that might interact with the nitrogen response). These are added as subtreatments. Though water is a critical determinant of response, it is not included as a treatment except in that it is recommended that the trials be conducted at several sites differing in water availability.

By comparing sites, water can be regarded as a variable. However, it is essential that rainfall is measured throughout the season and if possible that soil water availability is estimated at sowing. For this you will need to dry the soil samples at the research station in some form of oven. If soils are reasonably uniform, then three replicates of treatments will be good enough to give reliable results. If the soil is variable for various reasons, then consider four replicates.

Locations

Single cultivar nitrogen response trials, to determine if crops in the region will benefit from supplementary nitrogen fertilization, are ideally suited to integrating into farmer-sown fields. This has the time saving benefit of not having to sow the trials, the research will answer the question as to whether nitrogen is limiting yield, and a larger number of trials encompassing greater environmental diversity can be set up.

Nitrogen Treatments to Use

All N Applied at Sowing: the Basic Trial

The basic nitrogen fertilizer trial aims to determine if profitable responses to N fertilizer are possible in the farmers' crops if N is applied only at sowing. It has a similar aim to the starting point trial but has a more refined design.

A control plot and two or three nitrogen treatments all replicated three times, are necessary.

To determine the rates of N-fertilizer to be used (above that currently used by the farmer) multiply the average yield and protein of crops in the region by the correction factor as in the earlier budget table and round off the result to the nearest 10 kg N ha^{-1}. Amounts of nitrogen to be used as N-treatments can be this amount, half this amount and 50 percent more than this amount.

An example calculation for a crop of 2 t ha^{-1} grain with 11 percent protein would be: 2 t ha^{-1} grain x 11 percent protein x 2.34 = 52 kg N ha^{-1}.

This would be the pivotal treatment rounded to 50 kg N ha^{-1} and N amounts for all treatments become 0, 25, 50 and 75 kg N ha^{-1} additional to the farmer's normal application (*so the 0 treatment is the farmer's normal rate*). A further step below the farmer's normal rate can be introduced if nitrogen studies have not been carried out previously in the area. This will check whether current rates are economic.

If urea (46percent N) is the source of N then amounts of fertilizer to be applied are: 0, 54, 108 and 163 kg urea ha^{-1} (i.e. 0, 25, 50 and 75 divided by 0.46).

If ammonium nitrate (34 percent N) is the source of N then amounts of fertilizer to be applied are: 0, 74, 147 and 220 kg ammonium nitrate ha^{-1} (i.e. 0, 25, 50 and 75 divided by 0.34).

If topsoil has a neutral to alkaline pH then urea should not be applied to the surface, as there is a high chance of losses due to ammonia volatilization. Prior to sowing, urea can be drilled into the soil. When topdressing, nitrate-containing forms of nitrogen fertilizer should be used to minimize losses.

Layout of the Basic Trial

Below are a few rules to follow with this trial:

- locate the nitrogen plots in step fashion following the direction of the seeder; this avoids variability between plots due to inaccurately spaced tynes and allows the farmer to manage weeds in all plots with a single pass of machinery;
- make plots at least 2.5 m wide and 10 m long;
- mark out the plots initially with handfuls of lime or with sticks so the farmer can see them;
- use at least three replicates;
- at the five-leaf stage, mark out the plots by cutting paths around their borders;
- put brightly coloured corner posts around the trial area at the anthesis harvest. This will prevent accidental harvesting latr by the farmer;
- top-dress the nitrogen after sowing following rain. In alkaline soils preferably drill urea before sowing and then sow at 90°.

Timing of N Application: a More Complex Trial

The second level of trial complexity adds timing of N to amount applied. This level aims to manipulate the yield components to a more yield-efficient combination, but still without the farmer having to make decisions during the season.

The basic amounts of N applied remain as calculated above but they become three subtreatments: all applied at sowing as above; 50 percent applied at sowing and 50percent delayed until later in the season (beginning of stem elongation); and finally 100 percent delayed till later. The proportions and timing can be altered in line with local knowledge.

If the trials are being carried out at only one location and there is no consequent necessity to standardize the treatments for several farmers across a region, the crop stage for the delayed application can be decided to suit the season. In addition, more or less nitrogen can be applied based on how the season is progressing. This may be an additional subtreatment.

As a general rule, if the season is dry the application of supplementary nitrogen at stem elongation may result in haying-off and poor quality grain. However, if the season has been wet and there is adequate soil moisture, a positive response to more supplementary nitrogen is likely.

If N is top dressed during the season then the highest efficiencies of crop uptake are achieved if the nitrogen is applied just prior to rainfall. Rain takes the N into the soil quickly for root uptake.

Allowing for the Interactions of N with Other Nutrients: a Very Complex Study

If it is suspected because of crop symptoms or tissue analysis or the use of indicator crops that other nutrients such as phosphorus (P), potassium (K) or sulphur (S) are limiting yield, then these can be added as separate treatments. In the simplest study they would only be included as a subtreatment within the lowest N treatment (25 kg N ha^{-1} in the previous example).

Application of N-fertilizer may induce a nutrient deficiency through its stimulation of growth. If it seems that a trace element deficiency may be co-limiting yield then test this by including a treatment of a foliar application of a complete trace element mix, again within the lowest N treatment (25 kg N ha^{-1} in previous example).

If you get a response to the complete trace element spray then in later trials you can apply treatments with single elements to determine which of the trace elements is needed.

Measurements During Your Study

As emphasized earlier, the impact of nitrogen on crop growth and yield is dependent on the amount of water available as the crop progresses through its developmental stages. Consequently, to be able to interpret the effects of nitrogen correctly requires knowledge of soil water status.

Measurements of Crop Material to Determine Yield and Yield Components

Most of the measurements that follow are designed to help interpret the results of the trials. If you and the farmers are concerned solely with recording the responses to nitrogen without understanding in detail why they happened, the two critical measurements to take at maturity are plot grain yield and harvest index. Also take 100-grain weights from the grain threshed from the harvest index sample, for comparing kernel weights across treatments. Record this information for each plot, together with the plot identifier and date, on a table that was prepared in advance of harvest time.

For these less detailed measurements it is still vital to measure and record the plot areas actually harvested to work out production per unit area. Otherwise the comparisons between treatments become nonsense. It is also vital to bring all crop samples to similar water content before making comparisons, as water is very heavy. Quickly drying all samples on a hot day in the sun at the same time before weighing may achieve this satisfactorily. The sooner this is done after harvest the better, as any losses of material to respiration or accidents will negate the work of the studies.

You will need to take:

1. A hand harvest at anthesis if you are doing an in depth analysis of crop response;
2. A similar hand harvest will be required at maturity; and
3. Afull plot machine harvest at maturity.

When You Have the Numbers, How do you Interpret Them?

The following imaginary study assumes you have applied four nitrogen treatments (0, 25, 50 and 75 kg ha^{-1}, 0 is the farmer's normal practice) at sowing, following the design of the basic trial, and have duplicated the study over two farms. The primary aim is to determine whether farmers would gain an economic benefit from applying more nitrogen and if so, how that benefit had arisen.

Crop grab samples were taken at anthesis and again at maturity for yield component analysis and each complete plot without its end and edge border rows was harvested at maturity for plot yield.

The Anthesis Harvest

N-fertilizer kg/ha	0	25	50	75
Farm 1				
Biomass (t/ha)	6.80	7.80	8.21	8.40
Crop height (cm)	80	83	84	85
Farm 2				
Biomass (t/ha)	6.90	7.20	7.35	7.30
Crop height (cm)	77	79	79	77

What Does the Anthesis Table Say?

About Farm 1

- There was a 1.6 t h a $^{-1}$ growth response (6.80-8.40 t ha^{-1}) to a 75 kg ha^{-1} increase in N;
- crop height increased with fertilizer level.

These observations together indicate that in this season, when the farmer's normal practice was followed, N limited growth up to flowering.

About Farm 2

- N-fertilizer increased growth by 0.4 t ha^{-1} the effect being almost saturated by an increment of 25 kg N ha^{-1};
- small increments of N-fertilizer increased crop height, but not larger increments.

On Farm 2 it seems that the farmer's normal practice is close to optimum prior to flowering. Water, not N, was probably limiting biomass production at higher rates of N. This is suggested from the crop height reduction at high N.

The Maturity Harvest

What do the maturity data say for Farm 1:

Check the table at the bottom of the page as you read.

- N-fertilizer increased yield overall by 0.70 t ha^{-1}, but the first 25 kg ha^{-1} increment of N produced the most benefit;
- N also increased biomass and spike density presumably via more tillering, but a greater proportion of these additional shoots were sterile at high nitrogen (10percent versus 6percent).

This indicates that adding N intensified intra-crop competition for resources on this farm;

- increases in N-fertilizer resulted in marginally taller crops. This indicates that conditions following flowering were satisfactory. That all crops also grew taller between anthesis and maturity supports this suggestion;
- indications are that yield increase due to N was primarily via an increase in spikes m^{-2}, (leading to more kernels), but also through an increase in the number of kernels per spike.

There was the commonly observed trade-off however that kernel weight and harvest index both decreased in response to N-fertilizer.

In summary, maturity data for Farm 1 indicate that in the farmer's normal system N is a limitation to growth (and eventually yield) prior to flowering.

	Farm 1				*Farm 2*			
N-fertilizer kg/ha	**0**	**25**	**50**	**75**	**0**	**25**	**50**	**75**
Grain yield (t/ha)	3.25	3.78	3.93	3.95	2.96	2.75	2.65	2.50
Biomass (t/ha)	7.73	9.30	9.83	10.13	8.5	8.6	8.55	8.6
Spikes per m^2	307	352	365	375	302	325	337	325
Sterile shoots/m^2	20	25	30	40	20	25	35	50
Harvest index (%)	42	41	40	39	35	32	31	29
Kernels per m^2	9154	10755	11555	12000	8988	9744	10023	10288
Kernels per spike	29.8	30.6	31.7	32.0	29.8	30.0	29.7	30.3
Kernel wt (mg)	35.5	35.1	34.0	32.9	32.9	28.2	26.4	24.3
Height (cm)	83	85	85	86	78	79	79	77

So, depending on the relative cost of fertilizer and the grain price, adding further N could be beneficial.

Almost certainly the increment in yield of 530 kg ha^{-1} for 25 kg N ha^{-1} (just over one bag urea) would be an economic proposition.

However, a prospective further increment of 680 kg grain associated with two to three bags of urea may not be worth the parallel risk of haying-off.

What do the Maturity Data Say for Farm 2?

- N-fertilizer actually reduced yield by 0.46 t ha^{-1} and there was no response in biomass production;
- more spikes were produced with additional N, but not progressively at higher increments;
- crop height was unaffected by changed N and plants failed to grow taller between anthesis and maturity;

- more sterile stems at high nitrogen (13versus 6percent at low N) indicates that competition for resources increased with N on this farm;
- the yield decrease resulted primarily from a reduction in kernel weight though kernel number increased.

On Farm 2 it seems that it was fruitless to apply more N than in the farmer's standard regime as growth was primarily limited by lack of water both prior to and after flowering. Though water limitation prior to flowering was severe enough to halt height growth it did not limit floret fertility. Consequently, there could still be arguments for delayed N applications in seasons with late rain.

In the example, season late drought was so severe that when higher increments of N were applied, kernels were unable to fill even to 30 mg and harvest index fell dramatically. In effect the additional N was harmful for the crop.

Which Treatments Produced a Profit?

Gross margins calculations help you to make decisions about whether N applications are worth the cost and effort. They have been worked out for the two farms of this imaginary study. Your prices will differ from these examples quoted in A$ and based on costs in Australia, but trends will remain the same. Furthermore, your grain receiver may not be concerned with grain quality (protein content or screenings), only with weight. Re-work the examples in accordance with local regulations.

Assumptions in the Calculations

- That N fertilizer cost A$1 per kg applied and that uptake efficiency was 50 percent. Harvest index for nitrogen was 0.78;
- crops under farmer practice ('0' treatment) had 10 percent grain protein with a selling price of A$240/t when there are 0 percent screenings;
- grain price increases with protein and decreases with increased screenings (pinched and shrivelled grains passing a 2 mm slotted screen);
- the maximum tolerance of screenings is 10percent. Above this grain is unacceptable for milling and is deemed feed quality, reduced to A$150/t.

The Gross Margins

Farm 1				
N-fertilizer kg/ha	0	25	50	75
Grain yield (t/ha)	3.25	3.78	3.93	3.95
Input costs ($/ha)	150	175	200	225
Grain protein (%)	10.0	10.1	11.1	13.6
Screenings (%)	2.1	2.1	3.0	4.5
Value ($/t)	208	208	213	244
Crop value ($/ha)	676	786	837	964
Gross margin ($/ha)	526	611	637	739
$ Return per $N		3.4	2.2	2.8

For Farm 1 the gross margin per hectare increased with increasing amounts of supplementary N-fertilizer. However the return on money spent on N fertilizer was greatest at the low rate of N being A$3.4 for every A$1 invested.

Farm 2				
N-fertilizer kg/ha	0	25	50	75
Grain yield (t/ha)	2.96	2.75	2.65	2.50
Input costs ($/ha)	150	175	200	225
Grain protein (%)	10.0	12.7	15.3	18.5
Screenings (%)	3.0	6.5	9.0	11.0
Value ($/t)	206	217	240	150
Crop value ($/ha)	610	597	636	375
Gross margin ($/ha)	460	422	436	150
$ Return/$N		-1.5	-0.5	-4.1

For Farm 2 the greatest gross margin was for the farmer practice, the '0' treatment. Additional N-fertilizer wasted money especially at the highest N rate due to haying-off and associated unacceptable levels of small grains screenings.

Some Conclusions

The prime conclusion from these trials is that nitrogen is an essential plant nutrient that positively impacts on growth and yield when used thoughtfully.

However, it must be used in balance with other potential limitations to production, particularly water.

If farm management does not first attend to the basics of good seedling establishment with optimum plant populations, with the best variety sown at the optimum time, with adequate control of weeds and disease, nitrogen applications may provide no economic benefits.

The on-farm trials suggested here provide an opportunity for researchers to show farmers how to budget for a correct nitrogen balance, to work with them, to monitor water availability initially and through the season and match nitrogen and water inputs for optimum productivity.

They provide the tools to actually calculate for the farmer's unique circumstances, what these N-requirements might be.

They concomitantly provide an opportunity to see what the yield potential of a farm actually is and what the farmer might realistically aim for depending on the season.

5

Water Requirement of Different Crops

Climate, soil and water are the three basic resources which determine the nature of crops that can be grown successfully in a particular region. An efficient utilization of these resources is essential for optimum production of food and fibre for human life, feed for cattle and raw materials for industry. Climate determines the suitability of a region as a habitat for different flora and fauna, also the availability of water for production of Crops and other uses. Under a given set of environmental conditions production of crops is limited by the availability of nutrients and water. Soil provides anchorage for the plants and serves as a reservoir of water and nutrient required by them. Chemical fertilizer can supplement nutrients to poor soil, but there is no substitute of water for production of crops. As the resources are limited an efficient management is of utmost importance for sustaining and increasing agricultural production. Competing demands of water for domestic use, sanitation, industrial and recreational purposes makes it more essential to maximize the efficiency of water for agricultural production.

Agriculture consumes the largest amount of available water, yet, it uses water less efficiently. If food crisis is to be solved, there is no other alternative than to increase area under irrigation which can be made possible only when we use our present water resources most judiciously for irrigating our agricultural crops.

Soil, Plant and Water

Soil acts as a reservoir of water for the plants to use through root zone system. When rainfall is inadequate or untimely, application of irrigation water becomes essential for successful growth of crops. The factors influencing plant growth and productivity may be grouped as follows.

Soil Factors: The major soil factors are: soil moisture, texture, structure, density, salinity, fertility, aeration, temperature and drainage.

Plant Factors: The major plant factors are: types of crops, density and depth of rooting, rate of root growth, aerodynamic roughness of the crop, drought tolerance and varietal effects.

Climate factors: The climatic factors are: sunshine, temperature, humidity, wind and rainfall.

Miscellaneous Factors: Some of the major miscellaneous factors are: soil volume, plant spacing, crop and soil management.

Water is retained around the soil particles and in capillary pores by the forces arising out of surface tension and presence of salts. A plant has to overcome retentive forces to absorb water ; from the soil. The upper limit of water retention is limited to the field capacity of the soil. Any amount of water added to soil in excess of field capacity is not going to be retained in the root zone of the crop. When the soil texture is more of the clay content, the field capacity value increases. Time required for draining of the excess water also depends on soil texture. When the amount of water is reduced to a permanent wilting point, at this point or below plants can no longer absorb water and survive. The moisture content between field capacity and permanent wilting point is known as available soil moisture regime for plant growth. Humid conditions enables plants to draw soil moisture almost up to the permanent wilting point. On the other hand in hot and dry conditions, plants may show wilting symptoms in spite of enough available water present in the root zone.

Plants use only about 5% of the total water absorbed for physiological functioning. Most of the water is lost in transpiration and, therefore, it has to be replaced. Water absorption, conduction, translocation and water loss through transpiration are some of the main processes in plant growth. Water absorption by roots is dependent on the supply of water at the root surface and growth of roots in the soil mass. As the soil dries from saturated state, the rate of water movement in the soil decreases rapidly. The root system must expand continuously to provide the plant with sufficient water to replace the transpiration losses through the leaves. The internal water balance in a plant depends on the relative rates of water absorption and loss.

Water stress whether mild, moderate or severe can affect photosynthesis, respiration, growth and reproduction of plant. Water stress at certain critical stages causes more injury to plant growth and productivity. Some plants survive water stress owing to their drought resistant qualities.

Since field capacity and wilting percentage are peculiar to soils, water absorption may be effectively controlled by adopting an effective root zone. Shallow rooted plants like paddy and potato require frequent water application than moderately deep and very deep rooted crops such as wheat, groundnut, maize, sorghum, cotton and sugarcane. Moisture is extracted to a larger extent from the first quarter of the root zone than others.

Transpiration is the process by which water vapour from plant leaves enters the atmosphere. Though it is essentially a process of evaporation, it is different from the latter in several aspects. Evaporation from leaves meets with considerable amount of resistance like an open surface. Evapo-transpiration (En is the quantity of water transpired by the plants during their growth, or retained in the plant tissues, plus the moisture evaporated from the surface of the soil and vegetation, i.e. the amount of water between field capacity and wilting point within the root zone become available for ET. This is nothing but the consumptive use (CU) of water by the plants, because water used for metabolic activities is very insignificant (less than 1 % of En and, therefore, it denotes use for losses due to ET. Under field conditions, solar radiation supplies the energy for ET process. Wind is also important in removing water vapour from the field.

Use of Water by Plants

The kind and the extent of various plant roots largely determine the amount of water required at each irrigation. Roots usually; represent one half or even more of the total weight of the plants.

Water is absorbed mainly through root hairs of the roots near the terminal portion. Root hairs are usually very minute and they can not be seen by naked eye. Several hundred of root hairs may be located on a square millimeter of the root surface. With rapid root growth larger number of root hairs are sent into small soil capillaries and crevices between soil particles contacting and absorbing soil water.

After absorption by the roots, the water moves up the stem of the plant and into the leaves where it is given off as vapour. This loss of water is controlled to a limited extent by openings (stomata) in the leaves. The stomata are capable of opening and closing. There is very close relationship in the opening of stomata, transpiration rate and the soil moisture condition.

Water Requirement of Crops

The Water Requirement (WR) of crops depends upon retention and transmissivity of water in soil, absorption and transmission within plant, transpiration, effective rainfall, vapour pressure, and energy. WR is that quantity of water regardless of its sources required by a crop in a given period of time for its maturity. It includes losses due to ET or CU plus the losses during the application of irrigation water which may be unavoidable.

Irrigation requirement (IR) of a farm is the sum total of irrigation need for an individual crop in a specified time plus the losses occurring in field distribution such as seepage, percolation etc. Similarly, IR for a command area will constitute the sum of water needs for individual farms plus the loss taking place in the distribution system in that area.

Net irrigation requirement (NIR) is the depth of irrigation water, exclusive of precipitation, carry-over soil moisture or ground water contribution which is required for plant growth. It is that amount of irrigation water which is required to bring the soil moisture of the effective root zone to field capacity (PC). Thus, it is the difference between FC and soil moisture content in the root zone before irrigation.

Irrigation frequency depends on CU of a crop and the amount of available moisture in root zone. Sandy soils must be irrigated more often than fine textured deep soils. In general, irrigation should be given when, about 50 to 60% of the available soil moisture has been used from the root zone. The critical periods during the stages of growth are very important factor. Irrigation period is the number of days that can be allowed for applying one irrigation to a given area during the peak CU period.

The aim of the irrigation schedule should be to obtain maximum yield per unit area. In case of water resources being defined the objective should be to obtain maximum production per unit of water.

In heavy soils, depletion of available moisture is at a slower rate and in a lighter soil it is at faster rate. For paddy, it is enough to impound 5 cm of water and recharge to the same level once in four days or when the field comes to a saturation point but before the formation of hairline cracks, whichever is earlier. For maize at 25%

depletion of water, irrigation frequency should be once in four days and six days in case of clay soils.

Cotton and groundnut requires irrigation once in 10 days in red and 15 days in clay soils when the depletion of available soil moisture is 75 per cent. Other soils which fall in between red and clay loams, may have to be irrigated once in 10 or 15 days at the level of depletion.

Guidelines to Efficient Use of Water

The following are some guidelines for improving irrigation practices and efficiencies for different field crops.

1. Before irrigating, check the soil moisture in the root zone at several locations and estimate the amount of water require to bring the soil to field capacity. About 2 -3 days after an irrigation, check the soil moisture again. The moisture should be close to field capacity throughout the rooting depth. There should be no dry spots or dry layers in the field.
2. Determine the depth of water in centimeters applied to the field during irrigation. For this, the measurement of the stream size and the period of time for which water was delivered to the field are required. Then calculate the depth applied. For sprinkler irrigation multiply the application rate in millimeters per hour by the length of time in hours.
3. How does the estimation of the amount of water needed compare with the amount delivered to the field? About what efficiency irrigation was obtained? Note that high efficiency can be secured, but a poor irrigation may result if only a small amount of water is applied on the dry soil.
4. During irrigation, see whether the intake opportunity time is about the same throughout the field. When irrigation is done by the border method, does the water-stand about as long at the lower and middle of the field as it does at the upper end?

If furrows are used, does the water reach the lower end at I about one fourth of the total time that it is on the upper end? Are basins and level borders filled quickly?

5. Observe the amount of irrigation water flowing out of the field as waste. A large amount of surface water flowing out from a border indicates that the stream is too large or water has been running into the border strip for a long time. When the water in a well designed border (not a level border) approaches the lower end of the strip, the stream may be reduced or cut

off at the upper end. In this way, an even distribution will be obtained with little or no run-off. The stream size must be properly adjusted to the soil intake rate and to the border length if the border is to be evenly irrigated without excessive run-off. If furrow run-off is excessive, the furrow stream should be reduced to about one-third to two- thirds of the initial flow after the water reaches the lower end.

Irrigation Requirement of Crops

Several experiments on various crops for their irrigation requirements have highlighted the following results.

Paddy

Paddy is a semi-aquatic plant and covers about 35% of irrigated area in the country. Different varieties have been evolved to suit different regions, season and water availability. Cultural practices like puddling and transplanting reduce percolation losses, weed growth but increase the availability of plant nutrients and regulate soil and water temperatures. It improves photosynthesis in the lower leaves due to reflected light from the water surface. These operations may require about 200 -300 mm of water per hectare. Submergence below 50 mm for low land rice has been found to yield low. Recent researches have shown that continuous submergence throughout the growth period may not maximize the yield. Selective submergence during critical stages (initial tillering, panicle initiation to flowering) would be sufficient to maximize yield and to save water during the monsoon period.

However, during summer, continuous submergence has to be followed for maximum yields, Proper drainage helps to remove the toxic substances and regulate the oxygen supply to the roots. About 15 to 20 days prior to harvest, irrigations are stopped and water is drained to facilitate harvesting operations. A major problem in paddy irrigation is deep percolation losses which is 50 to 75% of water applied. Reduction of deep percolation of water is generally achieved by soil manipulation of three types viz. (i) Puddling (ii) compaction, and (iii) sub-surface placement of impermeable materials like bitumin and plastic films. Because of its prohibitive cost and mechanical difficulty, the third option is not practicable at farmers field.

Farmers generally continue to irrigate rice till about 4 -7 days before harvest. A recent study has stated that suspension of last irrigation for 14 -17 days before harvest causes more uniform ripening of the crop and economises 16 cm of irrigation water. However, termination of irrigation 3 weeks before harvest may cause marginal yield decrease, but it will save about 20 cm of irrigation water. The critical stages for irrigation in rice crop are tillering, panicle initiation, flowering and grain filling stage. The irrigation requirement of paddy at different locations.

Table: *Water and irrigation requirement or rice at different locations*

Sl.No.	Place	Water requirement (Mm)	Irrigation Requirement (mm)	Seasons
1	Kharagpur (WB)	18902150	1440N.A	July/Aug/-Nov/ Dec.Dec/Jan – Mar/ April
2	Cuttack (Orissa)	13001190	790780	June-SeptemberJan-April
3	Bhubaneshwar (Orissa)	14401650	7801630	June-SeptemberSept-Dec
4	Roorkee (U.P)	1620	750	June-October
5	Dhanauri (U.P.)	1630	910	June-October
6	Karnataka	1520	1170	June-October
7	Hyderabad (A.P)	N.A.	780	Mar-June
8	Coimbatore	N.A	1680	N.A. July/AugDec/ January
9	Chalakudy (Kerala)	N.A	1520	Feb-May
10	Delhi	2400	1600	June-October
11	Ludhiana	N.A.	1240	June-October

Wheat

Wheat is the second most important crop of the country and the area under this crop has been on the increase during the last two decades. It is grown during winter and about 59% of its area is irrigated. Recently introduced dwarf varieties are highly responsive to irrigation and fertilization. The critical stages for

Irrigation are crown root initiation, tillering, jointing, flowering, milk stage and dough stage. The crop develops a deep and dense root system unless restricted by an impeding layer in the profile. Therefore, it can effectively utilize profile-stored moisture, provided post -sowing irrigations are timed to encourage deeper proliferation of roots. The fact is especially important for the major wheat zone of northern India, where at the time of its sowing the root zone profile is either charged deeply with moisture from the proceeding monsoon or with pre-sowing irrigation. Generally, 4-6 irrigations are found to yield in maximum productivity at about 40 to 50% of depletion of available soil moisture. According to the availability of irrigation water, it may be scheduled.

Table: *Number or irrigations according to the availability*

Sl.No	***One***	***Two***	***Three***	***Four***	***Five***
1	CRI (21 DAS)	CRI (21 DAS)	CRI (21 DAS)	CRI (21 DAS)	CRI (21 DAS)
			Tillering (45 DAS)	Tillering (45 DAS)	Tillering (45 DAS)
2					Jointing (65 DAS)
3		Boot (75 DAS)	Boot (75 DAS)	Boot (75 DAS)	
4					Flowering (85 DAS)
5				Milk (100 DAS)	Milk (100 DAS)

Sorghum

Cultivation of sorghum is mostly confined to tropical and subtropical areas in Mabarashtra, Madhya Pradesh, Andbra Pradesh, Karnal Gujarat. Rajasthan, Uttar Pradesh, Tamil Nadu, Punjab and Haryana. Being the third important cereal of the country, the crop is planted in a total area of about 16 million hectares, of which 0.7 million hectares are irrigated. In Punjab, Haryana, Uttar Pradesh and some part of Rajasthan, it is mostly grown for forage rather than grain. The main growing season of the crop extends from March to October. In southern states, it is also taken as a rabi crop during October to February.

Experiments have shown that kharif crop do not require any irrigation, if there is sufficient rain during the season. It is a drought resistant crop and can withstand soil moisture depletion up to 75 per cent. Some. varieties of sorghum are of very short duration, adding to the adaptability, varied soil and water conditions. Seeding, flowering and sometimes milking are the critical stages when the crop is irrigated in case it is raised as irrigated crop.

Maize

Maize is grown both for grain and forage. Maize occupies an area of 5.9 million hectares. It is grown primarily as a kharif crop from March to October in Uttar Pradesh, Bihar, Madhya Pradesh, Punjab, Gujarat, Jammu and Kashmir, Himachal Pradesh, Orissa. Andhra

Pradesh, Karnataka, Maharashtra and Haryana. Some also come under rabi crop in the southern, mid west states and more recently in Bihar, Uttar Pradesh and Punjab. Preliminary studies have shown that rabi crop has a higher judicious water management and is important for improving the productivity of both kharif and rabi crops.

The crop has early vegetative, tasselling and silking stages as critical periods. After dough stage, there is no need of irrigation. The permissible depletion of soil moisture may be 25% in light soils and 50% in sandy loam to loam soils. Some research findings have revealed that crop should not be subjected to excess water- soil submergence during early growth. If the draining out of water in case of heavy rains is delayed, 30 to 60 kg N/ha may be added immediately following drainage to retrieve the yield loss.

Bajra

Bajra is an important millet grown during kharif in warm areas with a seasonal rain of about 15-20 centimeters. It is cultivated on an area of 11.7 million hectares mostly on relatively light textured soils of Rajasthan, Haryana, Punjab, Uttar Pradesh, Kamataka, Gujarat and Tamil Nadu. In most of the areas it is sown with the onset of monsoon and no post-sowing irrigation is applied. However, the crop, particularly the hybrid strains, has been found to respond to irrigation.

Flowering and milking stages are the critical stages of irrigation for bajra crop. The crop gives the optimum yield with irrigation based on 75% depletion of available soil moisture from the top 30 cm layer. Bajra is a drought resistant crop. If there is enough rain, kharif crop does not require irrigation. If there is no rain generally, two irrigations are required, first at flowering stage and second at the milking stage. If moisture is a limiting factor, irrigation should also be done at the time of ear head emergence because it is the most critical stage for moisture stress. Bajra does not tolerate water logging. So, do not allow rain water to remain in the field for more than a few hours. Proper arrangement for draining out of excess water must be made.

Barley

Barley is an important rabi cereal and is also used in brewery industry. It is grown on an area of only 1.75 million hectares in the states of Uttar Pradesh, Madhya Pradesh, Rajasthan, Bihar, Haryana, Punjab and Himachal Pradesh. Its cultivation is preferred to wheat under low fertility, low irrigation and late sown conditions. A few irrigation studies based on growth stages of barley have shown that barley shows favourable response to irrigation particularly if the rainfall is low. Generally, it require two to three irrigations to give good yield. One extra irrigation shall be required to sandy soils. If supply of water is inadequate, its efficiency should be increased by giving irrigation at critical stages of growth. If only one irrigation is available, it should be given near active tillering stage (30 -35 days after sowing). When two irrigations are available, one should be applied at active tillering and the other at flowering st3ge. On highly saline and sodic soils frequent light irrigation gives better result than few heavy irrigations.

Cotton

Cotton is cultivated primarily as a kharif crop on an area of 8 million hectares of which only 25% is irrigated. Major cotton growing st3tes are Maharashtra, Gujarat. Karnataka, Punjab, Madhya Pradesh, Andhra Pradesh, Rajasthan, Haryana and Tamil Nadu. The crop is sown in the hot months of April and May and is harvested during October to November in north-west and mid- west States and December to January in southern states. It requires more frequent irrigation in the southern states where the weather is warm and long season varieties are grown. In the North-west. cotton generally needs 2 to 5 irrigations depending upon the soil type and amount and distribution of seasonal rains. In drier and ill-distributed rainfall conditions, the crop needs 6 to8 irrigations with 50 to 75% soil moisture depletion.

Delayed irrigation prevents the plants from making excessive vegetative growth. The first irrigation should be given 40-45 days after sowing and subsequent irrigations should be light and be given at an interval of two to three weeks. The crop should not be allowed to suffer from water stress during flowering and fruiting period, other wise excessive shedding of flower buds and young bolls may occur resulting in the loss of yield. Cotton during its early growth is very sensitive to water stagnation for long periods. Therefore, proper drainage should be done.

Sugarcane

Sugarcane is an important crop of tropical areas. In India, it is grown on an area of about 3.19 m. ha. in Uttar Pradesh, Karnataka, Haryana. Bihar, Punjab, Gujarat and Madhya Pradesh. It is planted during January to March and is harvested after 10 -18 months in different parts of the country. It has a fibrous root system which can penetrate as deep as 2.5 m in well-drained soils devoid of any restricting layer and salinity. Thus, the crop utilizes most of the moisture stored in the root -zone. In the major sugarcane growing areas, 2 -3 months dry and hot pre-monsoon period of the growing season is very crucial from the irrigation point of view. During this period the young crop is adversely affected by severe soil moisture deficiency and excessively high soil temperature. In the rainy period, there is little need for irrigation unless the rains are too scanty and erratic. During post-monsoon growth period crop again needs irrigation adequately to meet its evapo-transpiration needs and in some cases to mitigate damage by frost.

The life cycle of sugarcane plant is divided into four distinct phases namely germination phase (from planting to 60th day); formative phase (from 60th to 130th day); grand growth phase (from 130th to 250th day) and maturity phase (250th to 365th day). The water requirement during formative phase and grand growth phase is maximum. Generally under north Indian condition, the water requirement of grand growth phase is met with rain, while the requirement of formative phase (during pre- monsoon period) has to be met through irrigation. The total water requirement for sugarcane varies from 200-300 centimeters. In northern India, autumn cane requires on an average 7 irrigations, 5 before monsoon and 2 after monsoon. Spring planted crops may be irrigated only six times. In drier climate and light soils, the crop may require 8 to 10 irrigations. In each irrigation, 3 acre inch of water should be applied. In Tarai areas, 2 to 3 irrigations before and one irrigation after monsoon are sufficient.

Drainage is also equally important in waterlogged areas. Drain excess water from the sugarcane field if they are flooded during the rainy season. Due to waterlogged condition, the quality of cane deteriorates greatly. Drainage greatly helps in increasing the yield and sucrose content of the cane.

Groundnut

Groundnut is the most important oilseed crop of the country. It is mainly grown in Gujarat, Andhra Pradesh. Tamil Nadu, Maharashtra, Karnataka, Madhya Pradesh, Uttar Pradesh, Orissa, Rajasthan and Punjab. Raised as kharif crop in warm areas of relatively high rainfall, it often suffers from periodic water deficits during long rainless intervals. This occasional moisture deficiency is one of the important factors contributing to low yield of kharif groundnut. Consequently provision of irrigation to groundnut would greatly aid in improving its yield. Even now the farmers of southern states are growing groundnut during rabi as an irrigated crop.

Being a rainy season crop, groundnut does not require irrigation. However, if dry spell occurs, irrigation may become necessary. One irrigation should be given at pod development stage. In the southern part of the country where groundnut is grown as rabi crop, 3-4 irrigations are necessary. Give the first irrigation at the start of flowering and subsequent irrigations whenever required during the fruiting period to encourage peg penetration and pod development. The last irrigation before harvesting will facilitate the full recovery of pods from the soil.

Mustard and Rapeseed

India is a major producer of mustard and rapeseed crops. These crops are primarily raised as rainfed crops during the rabi season on an area of 4.38 million hectares in Uttar Pradesh, Rajasthan, Madhya Pradesh, Assam, Haryana, West Bengal, Gujarat, Orissa, Punjab and BihaI. Due to low and uncertain rainfall during their growing season, these crops generally show favourable response to irrigations. It has been told that these crops need 1-4 irrigations, depending upon the soil moisture storage in the profile and the prevailing weather. Pre-bloom and pod filling stages are considered to be critical stages, therefore, irrigations at these stages are beneficial.

Gram

Gram is the most important pulse crop which is grown on an area of 3.5 million hectares mostly as a rainfed rabi crop in Madhya Pradesh, Rajasthan, Uttar Pradesh, Haryana, Maharashtra, Punjab, Karnataka etc. The crop has a deep tap root system and can make an efficient use of the profile-stored water in the well-drained alluvial soils of northern states and retentive clay or clay loam soils of mid-west and southern states. Although, water requirements of the crop are relatively low, it often suffers from periodic water deficits owing to low and uncertain rains in the rabi season. Consequently, the average yield of the rainfed crop is low. With recent advancement in irrigation facilities, a part of rainfed area under gram can be brought under irrigation for improving its yield. Therefore, there is need to conduct studies to determine irrigation requirements of the crop in various agro- climatic region.

If irrigation facilities are easily available, give a pre-sowing irrigation which will ensure proper germination and smooth crop growth. If winter rains fail, give one irrigation at pre-flowering stage and one at pod development stage. In no case first irrigation should be given earlier than four weeks after sowing. No irrigation should be given at flowering time. A light irrigaiton should always be given because heavy ones are harmful. Excess of irrigation enhances vegetative growth and depresses the yield.

Potato

Potato is an important vegetable crop which has two growing seasons viz. autumn and spring. Autumn potato is the main crop and sown in September-October and harvested in December- January. The crop requires frequent irrigations for optimum growth. A wet moisture regime is conducive not only for adequate water availability to the crop but also keeps soil strength low which permits better development of tubers.

Potato crop requires frequent and light irrigations at low moisture tension, irrespective of the varieties. This is particularly important with new varieties which produce large sized tubers. It is therefore important to keep the ridges moist but not too wet which may be attained by giving irrigation at regular intervals. Accordingly, the first irrigation should be light and is given 7 -10 days after planting or even earlier depending upon soil moisture content. the subsequent irrigation should be moderate to heavy which should cover 2/3rd to 3/4th height of the ridges. The intervals in subsequent irrigations could be about 15 days in heavy soils and 10 days in light soils. To complete its life cycle, the total water requirement of the crop about is 25 acre inch. Over-flooding results in soil compaction of the ridges which results into the poor aeration and poor tuber development. The crop should be given light irrigation if the frost is expected. The most critical stages for irrigation in potato are germination, stolon formation followed by earthing, tuber bulking which coincide 10 -12, 30 -35 and 55 -60 days after sowing. Moisture stress at these stages results in drastic reduction of tuber yields.

Irrigation requirement for different crops under various climatic conditions and soil types.

Table: *Irrigation requirement for different crops*

Sl.No	*Crop*	*Place*	*Soil type*	*Season*	*Irrig. Requirement*	
					Number	*Amt (mm)*
1	Wheat	Siruguppa(Karnataka)	Heavy black clay	Rabi	7	375
		Jobner(Raj.)	Loam sand	Rabi	6	405
		Hissar(Haryana)	Sandy loam	Rabi	7	420
		Ludhiana(Punjab)	Sandy loam	Rabi	5	360
		Roorkee(U.P)	Sandy loam	Rabi	3	220

Contd...

Sl.No	Crop	Place	Soil type	Season Irrig. Requirement Number	Amt (mm)
2 Maize	Delhi	Sandy loam	Kharif	2-3	100-150
	Hissar (Haryana)	Sandy loam	Kharif	5-6	300-360
	Arbhasi (Karnataka)	Clay loam	Kharif	3	150
	Siruguppa(Karnataka)	Black clay	Summer	10	510
	Bhavanisagar(T.N)	Loamy loam	Summer	25	1250
	Hyderabad (A.P)	Black clay	Kharif	2	120
3 Sorghum	Dharwar(Karnataka)	Loam to clay	Kharif	5	360
	Siruguppa(Karnataka)	Black clay	Summer	1-2	75-150
	Hyderabad(A.P)	Sandy loam	Rabi	4	300
	Coimbatore(T.N)	Clay loam	Summer	4	300
	Delhi	Sandy loam	Kharif	4	250
4 Bajra	Siruguppa(Karnataka)	Black clay	Kharif	2	150
	Hissar(Haryana)	Sandy loam	Kharif	3	200
	Delhi	Sandy loam	Kharif	2	150
	Anand(Gujarat)	Sandy loam	Summer	10	500
5 Groundnut	Dharwar(Karnataka)	Black clay	July-Oct	5	360
	Hissar(Haryana)	Sandy loam	July-Oct	4	300
	Chakuli(Orissa)	Loamy clay	Nov- April	-	690
	Jabalpur(M.P.)	Black clay	June- Oct	3	225
6 Cotton	Hissar	Sandy loam	May- Sept	4	50
	Delhi	Sandy loam	May- Sept	3	210
	Bhavani sagar	Red sandy loam	Feb- June	11	725
	Siruguppa	Clay	Aug- March	6	640
	Rahuri	Clay	July- Feb	2	150
	Hyderabad	Black clay	Sept- April	11	640
7 Sugarcane	Ludhiana	Sandy loam	Sept- April	8	620
	Roorkee	Sandy loam	March- Feb	11	660
	Karnal	Sandy loam	Dec- Jan	8	600
	Madhepura	Sandy loam	Dec- Jan	10	500
	Coimbatore	Clay loam	Feb- Dec	14	1000
	Nawasti	Clay	Dec- Jan	14	1750

Water Use Efficiency

Since crops require irrigation at different growth stages, the scheduling of irrigation should be planned very systematically. In the early stages, crops require usually lesser amount of water and any excess water applied is either wasted or propagated through more vegetative growth instead of deeper root penetration. More water is required in tillering, flower formation and fruiting stages. When the soil is alkaline or saline or if the water contains more soluble salts. water requirement is higher. Need of water varies according to different seasons. tillage practices. vegetative growth, cropping system and land shapes, therefore, climatic conditions, genetic variation in plants, agronomic practices, reduction of evapo-traspiration and water stress, fertilizer application, plant protection measures including weed control become the essential aspects in irrigation management. Since all these inputs determine yield productivity a proper synthesis among these may be necessary for an efficient system. Engineering concept expressed the efficiency in terms of net amount of water added to the root-zone or used in evapo-transpiration by a crop as a fraction of water diverted from some source. This would include different forms of water losses in conveyance and application. Therefore,

$$\text{Water use efficiency} = \frac{100 \text{ x Normal consumptive use of water}}{\text{Net amount of water depleted from root zone soil}}$$

This is the proportion between water delivered and that beneficially used. Since losses can occur during conveyance, application, storage and distribution, efficiency indexes could be measured during these different stages.

$$\text{Water conveyance efficiency} = 100 \text{ x} \frac{\text{Water delivered to the irrigated plot}}{\text{Water diverted from the source}}$$

$$\text{Water application efficiency} = 100 \text{ x} \frac{\text{Water stored in the root zone}}{\text{Water delivered to the field}}$$

$$\text{Water storage efficiency} = 100 \text{ x} \frac{\text{Water stored in the root zone}}{\text{Water needed in the root zone prior to irrigation}}$$

Water distribution efficiency = 100 x (1 -y/d)

Where,

d = average depth of water stored along the run

y = average numerical deviation from d

The aim of economic irrigation is to maximize the financial return per unit of water applied or amount of money invested in the irrigation projects.

$$\text{WUE} = \frac{\text{Crop yield}}{\text{ET}}$$

Where,

WUE = water use efficiency

ET = Evapo-transpiration (water loss)

Water Use Efficiency (WUE) is the ratio of crop yield to the amount of water depleted through evao-transpiration (ET). Water utilization by the crop is generally described in terms of kg of yield per hectare in millimetre. In the field, WUE would be the ratio of crop yield to total amount of water used. WUE could be increased either by increasing the crop yield or decreasing ET. Increasing crop protection is achieved through an integrated use of productive inputs. Decreasing ET requires adaptation of the plant varieties to the micro-environment and other genetic and climatic improvements. Optical irrigation and production efficiency should also be aimed in the economic evaluation of water use by crops.

The genetic variation in plants influence WUE. Those plants with higher rate of photosynthesis, usually have higher WUE, ego maize, sorghum, bajra, ragi and sugarcane. Most of the pulses, oilseeds and cereals like wheat, barley and oats have lower rates of photosynthesis as well as WUE. It is also found that the new hybrids and modem varieties have higher rate of WUE, both from agronomic and climatological considerations.

According to the WUE and by comparison among the different crops, one may be able to choose the crops to be cultivated.

Table: *Productivity or cereals per unit or water*

Sl.No	*Crop (new varieties)*	*Water requirement in a typical tract (mm)*	*Yield*	*Water use efficiency (WUE) per mm water*
1	Rice	1200	4500	3.7
2	Sorghum	500	4500	9.0
3	Bajra	500	4000	8.0
4	Maize	625	5000	8.0
5	Wheat	400	5000	12.5

Wheat has the highest productivity, followed by the millets and rice has the lowest water use efficiency (WUE). This may suggest the adoption of a non-paddy system, where all these crops are really competitive. But different regions have different cropping pattern and even similar cropping patterns may have different levels of productivity and water duty. Paddy is no doubt, a less productive user of water and any strict economic consideration may thwart its growth. The regional distribution may be such that paddy cultivation is specifically supported by climatic factors also.

Impact of Agriculutre on Environment

Agriculture is a vital sector within the economic structure. Coupled with the pressures for increasing productivity are growing concerns related to the availability of land, water and energy and the desire to preclude environmental quality deterioration. There are many concerns regarding environmental quality issue, including concerns about water, soil, air and noise. In this booklet an attempt has been made to high-light actual and or relative environmental impacts resulting from existing and emerging agricultural production technologies. The main aim and purpose is to provide information useful in precluding environmental deterioration while allowing for the necessary expansion of agricultural production.

Ever since, the emergence of green revolution, agriculture has become the most productive enterprise in the country, and the demands for increasing agricultural productivity are expected to continue due to an increasing population. However, the current problems and indicators of longer term concerns relative to agricultural productivity include the declining growth rate in agricultural productivity; shortages of resources like energy, water and soil and increasing concerns about deterioration of environmental quality due to pesticide and fertilizer usage and soil erosion. Hence, the need of the hour is to continue to produce an abundance of food and fibres to meet the growing demand of the population at reasonable prices and at the same time, the quality of environment and the scarce natural resources should also be conserved. For this new technologies must be developed to look into the agricultural production system.

Environmental Implications of Trends in Agriculture

Few environmental implications related to types of agricultural activities are being mentioned here.

Effects of Pollutants, from Agricultural Activities

Major pollutants from agricultural activities which include sediments, plant nutrients, heavy metals, salts, bio-degradable organics, pesticides, pathogens, odours and fugitive dusts, pesticide residues may contaminate surface waters, ground water, soil, the earth's atmosphere and thereby, affect the health of human beings through food chain. Pesticides reach water through direct surface run-off, ground water seepage, aerial drift during application and by being redeposited in waters upon volatilization. However, surface run-off pesticide concentrations depend upon the pesticide solubility, soil type, application techniques, quantity applied, and the amount and timing of rainfall.

It has been seen that the agricultural residues of nitrogen and phosphorus enter into the surface and ground waters from run off and leaching losses and from movement of sediments into surface waters, The total amounts of the nitrogen and phosphorus that can be lost from agricultural lands depends upon a number of factors. For instance, for the crop land, these include application rates, soil properties, terrain, soil erosion tendencies, crop management practices and the amount of rainfall. Soil sediments enter into water from all agricultural segments, and it is a transport agent of heavy metals, pesticides and plant nutrients. Any practice that increases or reduces sediment transport, affects the heavy metals transport accordingly. Slope and cover crop cultivation practices influence the amount of sediment lost. Soil salinity which results from irrigated crop practices, as well as the salinity acquired as inherent quality of the soil affects the quality of ground water and surface water, and soil productivity.

It is important to note that the potential environmental effects of pollutants from agriculture often cannot be assessed separately. Various other sources of pollutants lies within the common environmental receptors like, streams, rivers; lakes etc. Consequently, synergistic pollutant effects occur and associated environmental implications may result. By volume, sediment is the major pollutant in surface water, and it is also the transport agent for other residues. Sediments obstruct the drainage and irrigation canals, fills the reservoirs and lakes, and creates turbidity. Therefore, it is an economic issue, to clear the canals and reservoirs that have deposition of sediments. The surface and ground waters show increased salinity from irrigation practices. However, there is no danger to human health from increased salinity in surface waters. Its control also is primarily an economic issue since it involves industrial water treatment costs. High salinity levels can result in unpleasant taste and hardness, and a loss of aesthetic quality.

Salt build up in ground water can reduce crop yields, and crop production can become economically infeasible in areas of toxic salt levels. Increased levels of nitrogen com- pounds and phosphates in surface waters may lead to excessive algae growth which then increases the dissolved oxygen in water. The resultant stagnation in shallow water can cause increased mosquito populations and their consequent threat to health, a decrease in fish populations and other aquatic life, an overall decrease in animal and human water use and taste and odours. If there is a movement of ground water contain- ing high concentrations of nitrates into well waters, then is a major direct threat to humans from nitrates. In this reference, it is important to note that the US public health service has set 10 mg of nitrate expressed as N per litre as the upper safety limit. But many wells as well as some surface waters, exceed these limits. It is of primary concern that both persistent and non persistent pesticides may be found as residues in soil. Persistent pesticides may remain chemically active in the soil for several years, whereas non-persistent pesticides disappear within a few months of application. Primary concern is the uptake of soil pesticides by the crop which in turn can contaminate the foods.

Baseline Situation for Non-irrigated Crop Production

For assembling the basic information on non irrigated crop production, it is important to measure the soil erosion by estimating the average annual rates of soil movement from the crop land. Erosion losses from cropland may vary from negligible to more than 100 tons per acre. Nutrients are transported to surface waters largely by two means. One is nitrogen which is relatively soluble, reaches surface waters via run off and percolation. The other is phosphorus, relatively insoluble gets attached to sediment and enters the water through sediment movement. Although commercial fertilizer is the major source of nitrogen to crop lands for supplying over a million ton annually), animal wastes (manure) also contribute significant quantities of nitrogen. Animal wastes can pose more significant local N and P pollution problems than commercial fertilizers where per acre application rates are high. Since manure is disposed off on cropland in the vicinity of the feed lot) in which it was produced, the greater threat of pollution will be concentrated in the major livestock producing areas.

Even minor quantities of pesticide pollutant') create great concern because of their potential toxicity and persistence. Very little data! information exist on pesticide loading from crop land in surface waters. In addition to this, no estimate has been made about the total amount of pesticides entering the nations streams and rivers; however, investigations are being made with regard to run off water quality in local areas.

Environmental Implications of Trends in Non-irrigated Crop Production

The primary implications of environment on the trends in non-irrigated crop production must be known.

Crop Management Trends

The main trends that can be projected in crop management involve conservation tillage, crop sequence, and seed/plant improvement. In conservation tilling, the increased utilization of no tillage and reduced tillage practices will have the following major impact on both water and soil quality.

(i) The reduction in soil disturbance and the greater cover of residue on the cropland will impede soil erosion by reducing run off and this in turn will reduce the sedimentation of the surface water.

(ii) Although direct run off can be expected to decrease, percolation and leaching will increase since a greater part of the moisture will be retained on the cropland.

(iii) With the increased utilization of conservation tilling a greater infestation of cropland by insects and diseases will occur, as a result of this increased applications of pesticides will be required and will result in a greater potential for pesticides run off.

(iv) Although total run off will be reduced, the nitrogen concentration in that run off may be increased.

(v) Conservation tilling may also result in increase of nutrient run off both because of the large amounts of organic matter in the cropland and the possible greater applications of fertilizers.

(vi) The increased percolation will increase the potential for nitrates entering the groundwater and,

(vii) Due to the reduction in sedimentation, movement of phosphorus to surface water will be substantially decreased.

Hence, in the context the above mentioned points, the overall implications of conservation tillage will be a reduction of the potential for surface water pollution, however, the groundwater pollution potential will increase.

The trends in crop sequencing carry diverse implications. The mono-cropping and no meadow (meadow is a grassland especially used for growing hay) rotation practices would tend to increase water and land pollution. Mono-cropping increases the threat of insect and disease infestation, hence pesticidal applications may increase. In addition, mono cropping depending upon the type of crop involved, may increase or decrease erosion. Reducing the meadows from rotation sequence would increase erosion potential and typically require increased fertilizer application. Both of these would contribute to water and land pollution. Although crop sequencing practices would affect the environment adversely, relay and double cropping may have indirect beneficial effects. Both of these practices may impede erosion and reduce, in many cases, fertilizer requirements of a unit on output basis, i.e. hence, more intensive double cropping on one location may be preferred to extensive cropping on multiple areas. However, slight increases in insecticidal requirements would increase the potential for pesticide run off.

However, the environmental effects of seed/plant improvements are largely indirect. These genetic developments primarily affect crop yields and such increases have minor impacts on any given acre of cropland. The most significant implication stemming from increased yields would be a decrease in the cropland required to meet a specific level of demand. Although the overall cropland requirements will continue to increase, that increase would not be as great as it would have been without the improved crops.

Soil-water Management Trends

Trends in soil-water management include practices designed to reduce run-off and soil erosion, to enhance the moisture storage capacity of the cropland, and to decrease wind erosion. Erosion control-measures such as contour farming, terracing and use of winter cover crops are traditional methods of stabilizing the soil. Their use should be continued since, they are recognized to be environmentally sound and they benefit the producers. The principle feature of these practices is that they impede run-off and retard sediment movement and thus reduces both water and land pollution. A secondary effect is that they increase the moisture retention capacity of the soil and they increase percolation. This increased percolation results in a greater potential for nitrates entering the ground water.

Two trends that are significant for moisture conservation and water storage in soil are, fallow cropping and utilization of agents which reduce evapo-transpiration. Fallowing is common in semi-arid areas where rainfall is insufficient to produce a satisfactory annual crop. It not only promotes nitrification, but it aids in controlling noxious weeds. Additional benefits of fallowing are an increase of soil moisture and a slight decrease in the requirements for nitrogen fertilizer and herbicides. An adverse effect is the increased potential for both wind and soil erosion, particularly when fallowing is accomplished with cultivation. Evapo-transpiration reduction agents increase soil moisture and thereby significantly increase the crop yields in moisture deficient areas. The effect on the environment is similar to that of increasing crop yields through improved crops. The impact is indirect. Increased requirements for croplands are slightly reduced because of the greater production levels resulting from decreased evapo-transpiration.

Although wind erosion is most severe in semi-arid and arid regions under irrigation, it is significant in non-irrigated croplands. Reducing wind erosion not only stabilizes the soil and reduces land pollution, it also impedes sediments movement in active surface water. These effects also reduce the potential for air and surface water pollution.

Nutrient Management Trends

Nutrient management trends with environmental implications include new methods of applying fertilizers alternative nutrient sources, biological nitrogen fixation and certain technological developments. Alternative methods of commercial fertilizer application are generally more efficient and cause less soil disturbance during application. However, multiple applications are designed to apply fertilizers on cropland during favourable conditions of nutrient uptake by plants. However, multiple applications if increased even to a small extent may cause soil disturbances resulting in increased soil erosion. The major benefit is that the fertilizer remains in the soil a longer period of time before being utilized by the plants and is subject to greater risk .of loss through run off or leaching.

Biological nitrogen fixation development is a means of increasing nitrogen fixation by plant micro organisms, and methods of improving symbiotic relationships between plant and micro organisms. Also some genetic developments are anticipated in introducing nitrogen fixing capabilities into non- legume plants requiring high applications of fertilizers (eg. Com). With such nitrogen fixation developments considerable reductions in fertilizer usage would occur, and this would further reduce potential nitrogen run off and leaching. The collective trends in nutrient management (other than increasing the use of fertilizer) are expected to have an overall beneficial effect on the environment.

Pest Control Trends

Trends in pest control include improved methods of application, developments in new pesticides, resistant varieties and biological control. Improvement in the methods of applications are expected to have beneficial effects, e.g. improvements in aerial applications techniques will decrease the amount of pesticides applied to non-target areas, with a con- sequent reduction in pesticide usage and a greater efficiency of applications.

Development in dual application of fertilizers and pesticides will have both favourable and unfavourable environmental implications. Development of more resistant varieties are expected to reduce the requirements for pesticides. The most beneficial developments are anticipated in the area of new improved resistant varieties to pests such as insects, nematodes and birds, plus diseases. Biological control developments will also influence pest control practices. Potential developments involve the use of juvenile hormones, pheromones, sterile males, predators and parasites. Both beneficial and adverse environmental effects may occur with these developments. Biological control in some cases would decrease the requirements for pesticides on specific crops. However, this would reduce potential pollution problems, but the introduction of these biological control methods can have potential damaging effects on the environment if they affect the nontargeted plants or beneficial insects.

Baseline Situation for Irrigated Crop Production

It can be stated that significant pollution problems can occur in both surface water (streams and rivers) and ground water by pollutants from irrigated croplands. The major surface water pollutant (by volume) is sediment resulting from soil erosion. While soil erosion is not a major concern in irrigated crop land, the sediment produced can also carry other pollutants such as pesticides and fertilizers. Fertilizer pollution problems can be expected to occur in those geographical areas having considerable cropland acreage receiving high rates of application. Irrigation return flow salinity can be a significant pollutant to surface water and ground water and the composition of soil. The major components of salinity include water soluble compounds of the cations calcium, magnesium, sodium and of the anions carbonates, bicarbonates, sulphates and chlorides. Minor amounts of iron, aluminium, manganese and other cations are also present. Salinity may occur during normal irrigation practice also as a portion of the irrigation water is carried to the subsoil or ground water. This percolating water carries with it salts accumulated in the root zone. Some may be eventually collected by drains and returned to the main stream. The salt concentration of the irrigation return water may be more than that of the initial irrigation water. Another fraction accumulates on or near the surface. Salinity problems are associated principally with cropland irrigated with surface water.

Environmental Implications of Trends in Irrigated Crop Production

The points mentioned for non-irrigated cropland are also applicable to the trends that are seen in irrigated cropland management practices such as crop management, soil water management, nutrient management and pest control. Only the points relative to the irrigated cropland will be mentioned here.

Reduced water application during irrigation has both beneficial and adverse effects on the environment. The primary benefit involves the conservation of water resources. Adding excessive moisture during irrigation not only wastes water, which is scarce in supply, but also may damage the crop. On the other hand recycling may increase problems of salinity. However, direct monitoring of irrigation requirements is a pre-requisitor for the proper management of later practices concerning reduced irrigation.

Environmental Effects of Conservation Tillage Practices

Tillage refers to the preparation of land for planting. The purpose of this practice is to establish a weed free seed bed that permits the incorporation of fertilizer, uniform seed placement, quick germination and a good early growth rate. For years, the usual/standard tillage method followed has been to turn the soil completely along with residues just before planting. Tillage that completely inverts the soil and buries all crop residues generally will leave the land much more exposed to the erosive forces of wind and water than tillage that twins the soil-less and leaves more crop residue on the surface. This latter type of technology is called conservation tillage.

Conservation tillage is a system which covers a variety of tillage practices, but they all have three features in common.

i. They use some implement to prepare the seed bed.

ii. They leave enough crop residue on the soil surface to significantly reduce erosion.

iii. They rely less on cultivation and more on herbicides to control weeds than conventional tillage method.

Conservation tillage proves to be very effective in the control of wind and water erosion. Soil erosion by wind can be a problem wherever the following conditions of soil, vegetation and climate prevail. These conditions include; i) a soil that is loose, dry and reasonably divided; ii) a smooth soil surface on which vegetative cover is absent or sparse; iii) a large enough field; and iv) wind that is strong enough to move soil.

Conservation tillage is effective in wind and water erosion control since the presence of crop residues on the soil surface presents a barrier to wind and retards run off. The formation of larger soil clods that occurs during conservation tillage systems serves as a barrier to wind and water movement. This capacity to reduce erosion is one of the most important feature of conservation tillage technology.

Conservation tillage is found to be very effective in reducing sediment in run off water. This also has influence on pesticides in run-off water. Some pesticides either are not very soluble or they adhere tightly to soil particles. In these cases, erosion reduction prevents or greatly reduces entry of pesticides into surface waters. Thus, conservation tillage act to lessen the impact of such pesticides, and also' they reduce run off but do not eliminate it. Therefore, because conservation tillage has such an increased reliance on pesticides particularly herbicides, it is a greater threat to the environment than conventional tillage as far as pesticide damage is concerned. This is particularly possible in soil clays in wet tropical regions where such soils do not bind the herbicides to their surfaces. In such environments, a large portion of the herbicide can be carried into water bodies.

Environmental Effects of Organic Farming

Organic farming can be defined as a production system which avoids or largely excludes the use of chemical fertilizers, pesticides, growth regulators and livestock feed additives. To the maximum extent possible, organic farming approach derives maximum benefit through crop rotations, crop residues, animal manures, legumes, green manures, off farm organic wastes, mechanical cultivation and aspects of biological pest control. All these practices contribute towards maintaining soil productivity and tilth, supplying plant nutrients, and con- trolling insects, weeds and other pests.

Environmental benefits and opportunities that lend support to organic farming practices include reduced soil erosion, reduced nutrient and pesticide pollution, reduced energy requirements, increased recycling of organic wastes and enhanced soil and water conservation. These are mentioned here briefly.

Reduced Soil Erosion

Practices involving the use of cover crops, green manure crops and organic matter management all help to control soil erosion. However, the advantage of organic farming methods is for controlling soil erosion over that of chemical intensive systems.

Reduced Nutrient Pollution

Organic farmers avoid or restrict the use of commercial fertilizers and instead rely more on recycling of nutrients in their farming operations. This reduces the opportunity for residual nutrients in soil which might be subjected to leaching. Reduced soil erosion also minimizes nutrient transport from fields.

Reduced Pesticide Pollution

Organic farmers greatly restrict or avoid the use of pesticides in their operations. Decreased use of pesticides help to reduce run off if the agricultural chemicals in some areas and to reduce the spread of chemical residues in the environment.

Reduced Energy Requirements

Organic farmers use appreciably less total energy for producing most crops than conventional farmers. Considerable quantities of energy are saved on organic farms by the use of crop rotations and the application of organic wastes in place of chemical fertilizers especially nitrogen.

Increased Recycling of Organic Wastes

In addition to the recycling of on-farm organic wastes, some organic farmers utilize off-farm sources of organic wastes, including sewage sludge and processing wastes, thereby achieving a higher crop yield potential. This helps to resolve a disposal problem, while contributing greatly to improved soil productivity.

Enhanced Soil and Water Conservation

Organic farmers make extensive use of crop and soil management practices that effectively protect the soil and improve infiltration. Improved water conservation with organic farming may provide opportunities for farmers in dry-land areas.

Impact of Agriculture on Water and Soil

Technologies used for various agricultural activities can have significant impacts on both water (surface and ground water) and soil environments. Here an attempt is made for brief summarization of these impacts based on current practices used. Identification of water and soil impacts include information on nutrient pollution of surface and ground water, pesticide pollution of surface and ground water, sediment losses and soil erosion and general ground water concerns. Due to the importance of the soil environment relative to the nature and extent of both water and soil impacts, it is essential to first give a brief background information on this environmental media before actually dealing with the information regarding environmental impacts.

Nutrient Pollution of Surface and Ground Water

Soil contain large densities of bacteria that decompose soil organic materials and form an integral part of the soil organic composition.

Bacterial densities in soil range from several thousands to 200 million bacteria per gram of soil. Nitrogen and phosphorus resulting from fertilizer applications and other agricultural practices can contribute to both surface and ground water pollution. Surface water contributions are considered to occur from non-point source discharges which primarily occur during rainfall where run off from the land surface carries sediment, pathogens, sediment adsorbed chemicals, dissolved chemicals (such as nutrients and pesticides), heavy metals, and easily oxidizable organics into adjacent water ways. Dissolved chemicals may also percolate through the soil to interflow regions and or ground water, and be discharged in sub-surface flows. Nutrient pollution of ground water involves sub-surface transport to water bearing zones. The transport and fate of nitrogen in the sub-surface is dependent upon the form in which nitrogen enters and various biological conversions which may take place.

Nitrates (N°3) can be formed by nitrification involving ammonium ion conversion to nitrites and then to nitrates. Nitrification results from an aerobic reaction performed primarily by obligate autotrophic organisms and N03 is the predominant end product. Nitrification is dependent on the aeration of the soil, which in turn is dependent on soil characteristics, percolation rate, loading rate, distance to impervious strata and distance to ground water. However, complete nitrification may occur in more clayey soils such as silt loams and clays.

Denitrification is another important nitrogen transformation in the sub-surface environment. It is the only mechanism by which the N03 concentration in the percolating (and oxidised) irrigation water can be decreased. Denitrification, or the reduction of N03 to N2O or N2, is a biological process performed primarily by heterotrophs. In the absence of 02, N03 acts as an acceptor of electrons generated in the microbial decomposition of an energy source.

Based upon the forms of nitrogen in fertilizer applications, and the biological transformations which can occur in the sub-surface environment, there are two forms of major concern relative to ground water pollution ammonium ions (NH4) and nitrates (NO3). Ammonium ions can be introduced directly from fertilizer applications, or they can be generated within the upper layers of soil from the ammonification process (conversion of organic nitrogen to ammonical nitrogen). The transport and fate of ammonium ions may involve adsorption, cation exchange, incorporation into microbial biomass, or release in to the atmosphere in the gaseous form. Adsorption is probably the major mechanism of removal in the sub-surface environment

Pesticide Pollution of Surface and Ground Water

Pesticide usage has aided in increasing agricultural productivity but this potential may prove detrimental to health and environment. This include the destruction of non-target organisms; deposition of residues that magnify in food chains and eventually injure predatory animals at the top, including man; and direct health effects on pesticide users. The chief concern related to the impacts of pesticides on the natural environment is associated with the fact that pesticides can be transferred from their original application sites to other locations by erosion, drift run off and biological means.

The transport and fate of organic contaminants in the sub-surface environment is a relatively new area of concern, and hence enough literature regarding this is not available. A variety of possibilities exist for the movement of organics, including transport with the water phase, volatilization losses from the soil system, retention from the soil due to adsorption, incorporation, microbial or plant biomass, and bacterial degradation. However, the relative importance of these possibilities in a given situation is dependent upon the characteristics of the organic, the soil types and characteristics, and the sub surface environmental conditions.

Sediment Losses and Soil Erosion

As a result of heavy demand for an increase in agricultural production, there have been instances of inclusion of marginal lands into production and removal of soil erosion control structures. When this is combined with the use of heavy dose of chemical application, the stage may be set for soil erosion problems. Soil erosion has been of concern since it can cause de- crease in soil productivity and increase in water and air pollution. Soil erosion lowers soil productivity through the loss of storage capacity for plant available water, loss of plant nutrients, degradation of soil structure and decreased uniformity of soil conditions within a field. Loss of top soil nearly always reduces the inherent productive potential of a soil. The characteristics of individual soils, however can strongly affect the quantitative impact of soil erosion upon productivity.

Nutrient Management Practices

Nutrient losses through the run off water are primarily a result of fertilizing practices. The use of chemical fertilizers to supplement the nutrients supplied by the soil has long been recognized as necessary in most soils for optimizing crop yields and plant quality and reducing erosion by increasing vegetative cover. The potential for pollution from fertilizers will generally be highest where large acreage are treated with heavy doses of fertilizers. However, the nutrients are moved from agricultural land by leaching, direct run off and in association with sediment from erosion. A number of practices will reduce direct run off and erosion and, thus, reduce the nutrient transport. However, in some cases, such as leaching, additional or alternative practices will have to be used to achieve the desired degree of control. These practices involve changing the use of nutrients, for instance, these may include eliminating excessive fertilization, timing the nitrogen application, using crop rotations, using animal wastes for supplying nutrients instead of ploughing under green legume crops, incorporating surface applications, controlling surface applications and using legumes in haylands and pastures. .

Pesticide Management Practices

The use of chemicals to control the crop pests has been on an increase. Investigations reveal that some chemicals are highly toxic to fish or other aquatic fauna and can persist in the aquatic environment for a long time, so that even very low levels of these pesticides in run off may be of environmental concern. On the other hand, many agricultural chemicals are not acutely toxic to animal life, do not persist from one crop season to the next, and do not accumulate in food chain organisms; consequently they may be used at normal application rates without fear of causing unacceptable environmental damage.

Some of the methods that can be adopted to minimize the pesticide application in agricultural practices include:

i. adoption of production techniques which do not use pesticides;
ii. use of alternative pesticides that are not water soluble;
iii. optimization of pesticide formulation;
iv. reduction in excessive treatments;
v. optimization of the time of day for pesticide application;
vi. optimization of the dates of pesticide application
vii. use of optimum pesticide application rates; and
viii. management of aerial applications.

Besides all these measures, a reduction in run off or erosion will also reduce loss of applied pesticides, and practices that control run off and erosion should always be considered in pesticide pollution control.

Impact of Agriculture on air Quality

Agricu ltural practices can have significant impacts on the air environment. In addition to the impacts of agricultural technologies on air quality, which may largely result from non-agricultural activities, can exert undesirable effects upon agricultural crops and animals. These impacts of agriculture on air quality based upon the usage of some emerging agricultural technologies are in terms of tillage operations used, air pollutants obtained from open burning, from wind erosion, from unpaved roads, from agricultural vehicles used in fields, pollutants obtained from harvesting and grain handling and pollutants tram pesticide applications.

During a tilling operation, dust particles from the loosening and pulverization of the soil are induced into the atmosphere. Open burning of crop residues, as well as debris from land clearing operations, represents a potentially significant source of air pollutants from agricultural activities. Ground level open burning is affected by many variables, including wind, ambient temperature, composition and moisture content of the debris burned, and compactness of the pile. In general, it can be said that the relatively low temperatures associated with open burning increase the emissions of particles, carbon monooxide, and hydrocarbons and suppresses the emissions of nitrogen oxides. Particulate emissions represent the primary type of air pollutant resulting from wind erosion. Factors influencing particulate emissions from wind erosion include soil type, precipitation patterns, exposed area and wind speed.

Vehicles used for the agricultural operations include tractors, cars and trucks. The types of air pollutants emitted from such vehicles include carbon monoxide, hydro carbons, and nitrogen oxides. Harvesting and grain handling can produce large, quantities of particulate and hydrocarbon emissions. Further, substantial quantities of pesticides can become air-borne during and following the aerial spraying operations. The key mechanisms for pesticide introductions into the atmosphere include aerial drift and evaporation (volatilization) from soil and plant surfaces. Volatilization is a major pathway of loss of pesticides from soil, and loss is greater from moist than from dry soils. The relative amounts of a given pesticide entering the air and surviving to some downwind site depend on the nature of the source (eg., the type of application in agricultural operations), its physical properties and chemical reactivity, its form once it enters the air, and that which exist throughout the process. One concern pertaining to air borne pesticide residues is related to potential human health' effects, particularly for agricultural workers.

Air Pollution Effects on Crops

In order to understand the effects of air pollutants on agricultural crops, it is necessary to consider the basic structure of plant leaves and the functioning of the various components of the structure. The major plant processes that occur include photosynthesis, transpiration and respiration. Several components of the leaf structure are involved in the photosynthesis process. The leaf veins are involved in moisture and nutrient transport to the leaves from the soil and root system of the plant. The effects of air pollutants on crops must be considered, in terms of environmental factors including temperature extremes, excess water, water deficiency, nutrient deficiency and bacterial or viral disorders.

High temperatures may cause chlorosis which is reflected by yellowing of the leaves, while low temperatures may also cause chlorosis or necrosis. Excessive water can damage veins of the leaves, and lead to plasmolysis, while a water deficient conditions is reflected by necrosis. Nutrient deficiencies to plants can cause chlorosis or necrosis, while bacterial and viral disorders may be manifested by plasmolysis. In brief, it can be said that many factors influence the visible response of a given species of plant to a particular pollutant. Among the more important are the age and variety of the plant itself, the concentration of the pollutant, the length of exposure, the vigour of the plant and growing conditions before, during and after exposure. Thus, the environmental factors can act as potentiators for the specific air pollutants in terms of effect, or they can cause some typical symptoms, specific to the type of the pollutant.

Air Pollution Effects on Animals

Under both acute as well as chronic conditions, air pollutants can affect the animals significantly. In terms of chronic effects, the air pollutants which have received the utmost attention include fluorides, arsenic and lead. Air borne fluorides have caused more world wide damage to domestic as well as farm animals than any other air pollutant. The animals most affected by fluorides include cattle and sheep, and fluoride damage to animals have been of great concern. The symptoms of fluorosis in cattle are a function of whether or not there is an acute exposure or a chronic exposure. In general, the acute symptoms may include lameness, stiffness, lack of appetite and thirst, diarrhoea, muscular weakness and possibly death.

The chronic symptoms may include skeletal changes, lethargy, emaciation, poor health, and possibly a poor reproductive efficiency.

Fluorosis can result from drinking water with high fluoride contents which may affect the human being as well as animals. However, this is possible that in some cases diseases and symptoms as indicated above, can occur from factors other than exposure to atmospheric fluorides. Some other examples of air pollutants which have exhibited effects of domestic animals include ammonia, carbon monoxide, hydrogen sulfide, sulphur dioxide and nitrogen oxides. There is some evidence that ammonia can have harmful effects on poultry, carbon monoxide can affect quite a number of animals and dusts can also affect rabbits, hydrogen sulfide can affect poultry, sulphur dioxide can affect cattle and sheep and nitrogen oxides may also affect quite a number of farm animals.

Soil Fertility and Grasses

Soil fertility is measured by the capacity of the soil to produce crops. It is dependent on its ability to supply to the roots of the growing plant with water, oxygen and nutrients. The term fertility should not be confused with productivity which depends not only on the inherent fertility of the soil but also on the way it is managed. Productivity provides a measure of soil fertility only under a given system of utilization. In systems of alternate cropping it is the productivity of the whole crop rotation that is important. Pastures and fodder crops will be productive only if they are fed to productive livestock.

While chemical fertilizer is essential the physical condition of the soil is equally important. Soil organic matter, by itself is not necessary for plant growth, since plants can grow perfectly well in solutions of chemical nutrients. But it is required for stabilizing the soil crumb structure and increasing its ability to retain moisture in soils. Organic matter also increases ion exchange capacity markedly. The breakdown of the soil organic matter by the soil flora and fauna provides a source of gradually available nutrient" to plant roots. In the absence of. this biological activity, organic matter accumulates in the form of peat or "raw" humus, and the chemical nutrients stored up in it are not available to plant roots.

When virgin f9rest or grassland is cleared for agriculture there is a rapid loss of organic matter and structure, until eventually a new equilibrium is reached depending on the climate, soil type, and system of agriculture. In humid- temperate regions, this new equilibrium is more stable but under hotter and drier conditions, the new equilibrium may be far below a safe level and cannot be maintained even with liberal applications of organic manures.

Soil Fertility Status in Tropical Areas

In tropical areas, stable civilizations have in the past developed only in areas where agriculture has been based on the cultivation of irrigated rice. The population in many parts of the tropics is still dependent on a shifting form of agriculture. After a few years of cultivation, land must be allowed to revert to natural grassland or bush for considerable periods to allow fertility to be restored. Under these conditions, economic and social advance is difficult.

Increasing population and the demand for a high standard of living, increased the intensive cultivation of cash crops. This intensive cultivation caused a gradual decrease in soil fertility in many tropical and subtropical areas. In the more densely populated areas, continuous cropping of food and cash crops has in many places resulted in serious soil deterioration and reduction of yields. We need to develop stable systems of farming which will help to increase and also maintain the levels of production. Integration of animal and crop husbandry in systems of mixed with the return of animal manures and crop residues to the soil improves soil fertility. Productive pasture or fodder crop too can maintain soil fertility. The increased use of legumes in cereal rotations as green manures, forage crops or as edible pulse crops also offers yet another valuable possibility for improving soil fertility. Deep rooted grasses produce large quantities of root organic matter and increase the ability of the soil to absorb rainfall, and leave the soil in a condition which is resistant to erosion.

Introduction of conservative farming may retard soil erosion. But they cannot by themselves, arrest the gradual depletion of soil nutrients which takes place as a result of the export of crop and animal products from the land. Soils in the tropics are in general, poor in unweathered minerals from which the store of plant nutrients may be replenished. The use of inorganic fertilizers on a large scale is rapidly becoming essential on all but the most fertile soils, not only replace nutrients removed in crop and animal products but also as a necessary basis for the introduction of fertility building components.

Livestock can act as a means by which fertility is transferred from the nature grazing lands to the cultivated land. But often the animal manure and especially the urine is wasted on the village grounds and the dung is used as a source of domestic fuel. The return to the land of all available waste products, either direct or through the animal is most important. Products such as oilseeds cakes should be fed to animals in the place of origin instead of being exported. By this the mineral nutrients may be returned to the land through the manure.

Tropical dry areas are poor in fertility mid much management practices are needed to maintain the fertility of the soil. The main reason for the poor fertility is the moisture shortage and organic matter content in the soil. The fertility maintenance consists mainly in building up soil moisture and organic matter.

Fertility Status in Humid Temperate Regions

In humid temperate countries conservative systems of farming based on alternate crops and the use of fertilizers are successful in maintaining high yields from arable lands. Different systems have been developed depending on the agro-climatic conditions and on the general farming back ground. These systems may be based on the grazed grass/legume with direct return of dung and urine to the soil or growing fodder crops or rotation hay crops with careful conservation of animal manure for application to the arable crops, or permanent pasture and hay crops with high rates of fertilization and full use of animal manure or on a mixture of all three. These systems of alternate cropping are all profitable because they are usually associated with a profitable animal industry. It will be seen that different soil conditions call for different methods of cropping and it is not generally possible to say that one is better than another. All can be successful in maintaining high arable yields. The legume, as a source of nitrogen, is important for maintaining soil fertility. Pure grass crops deplete the soil of available nitrogen and can remain productive and build up fertility only if large dressings of nitrogenous fertilizers are applied. Farmers in humid temperate countries are fortunate in having a large number of productive legumes at their disposal. Humid temperate regions are rich in organic matter and soil moisture and hence soil fertility maintenance is not a major concern.

Fertility Status in Temperate Semi-arid Conditions

Under temperate semi-arid conditions the soil moisture is the limiting factor. Perenniel grasses have been found to deplete soil moisture and to decrease the yield of the following crop. No good perennial legume is available for inclusion in rotation pastures and the build-up of fertility under pure grasses is very slow. Continuous clean cultivation, particularly on the lighter soils, has in some cases resulted in serious wind erosion and it has been found necessary to sow such areas down to permanent or semi permanent grass land. It seems that, under these conditions, the use of inorganic fertilizers combined with the incorporations of all the crop residues in the surface of the soil by stubble mulching may be sufficient to maintain fertility indefinitely and prevent serious erosion except perhaps on very sandy soils. Short periods under grass are of no value.

Fertility Status under Sub-tropical Conditions

Under subtropical conditions grazed lands based on annual self regenerating legumes and the use of phosphates have been successfully introduced into cereal rotations in areas with winter rainfall. This may not possible when the holdings are small and fragmented. There is however, a great need for fodder crops to provide more feed for livestock and to raise the low level of animal production. Experiments have shown that inclusion of fodder or grain legumes in the cereal rotation can lead to a marked increase in overall productivity. In the drier areas, stubble mulching is advantageous. But this would require more machinery than is generally available on the present small holdings. It would reduce pressure on stubble grazing and straw which at present forms the sole fodder contribution of the arable land.

The maintenance of soil fertility is perhaps more difficult under sub tropical summer rainfall conditions. So far no satisfactory perennial legume has been found for inclusion in grazed tracts in areas with a long dry season. Experiments in South Africa and Southern Rhodesia have now shown that pure grass lands can build up fertility to a high level if sufficient nitrogen is given. In these areas there is now a possibility of creating a prosperous farming industry based on milk or beef production. However, introduction of such systems is more difficult both at the large and small farm level.

Maintenance of Soil Fertility

In the tropics, except in the areas devoted to plantation agriculture and a few non-typical areas at high altitudes, holdings are generally very small and are cultivated by hand or with primitive animal drawn implements. In the densely populated areas, continuous cropping over the centuries has often resulted in serious soil deterioration with yields reduced to low levels. In the more thinly populated areas, shifting cultivation is still very widely practised. Provided the resting period under bush or grass is sufficiently long, crop yields can be maintained at a reasonably high level; the overall productivity of the land is, however, low as the resting period under bush or grass is totally unproductive. With increasing population and increasing cultivation of cash crops, it is becoming necessary to try to find ways in which this overall productivity can be increased.

Deep-rooted perennial grasses have been found useful in building up fertility in the tropics without application of fertilizer. This is true particularly in areas with moderate rain fall (75-100 cm). There is a marked improvement in soil structure and in the capacity of a soil to absorb rainfall after ploughing out a grass fallow. This improvement is, however, short-lived. The main beneficial effect of the grasses lies in their ability to recover plant nutrients, including nitrates, from the sub-soil. This is possible by including perennial legume with the grass. The legume can be a tree too. It builds up organic matter and nitrogen in the soil. The grasses also will benefit from the legume.

Leaving land fallow with grasses is probably more useful for soils containing not less than 25 per cent of clay. On sandy soils, particularly in areas of heavy rainfall where nutrients are leached to a much greater depth, fallowing with bushes is more effective. On suitable soils, rotations of three year grass followed by three years cropping have given good results in a number of places. This marks a distinct advance on the bush fallow where three years of cropping may have to be followed by a resting period of six to twenty years or more under bush, depending on soil and climatic conditions.

There is the further possibility that the grass fallow may be utilized for grazing. Preliminary studies in Uganda have shown that grazing of the grass fallow did not reduce yields of the following crops, although compaction by trampling reduced the absorption of rain water by the surface soil. Careful grazing management will be essential to reduce transplanting and compaction of soil. Overgrazed, heavily trampled grass lands are not useful in building up fertility and are subject to serious erosion on sloping land.

In tropical Areas

During the rainy season there is ample food for livestock in most tropical countries. Generally considerable uncultivable areas which provide useful grazing are there in tropical areas. Weeds growing by the roadside and among the crops also provide palatable fodder. During the dry season, there is always an acute shortage of quality fodder. But sown grass or grass/legume cropping will be required mainly for cutting and conservation as hay or silage. Such cropping will require fertilizers for production and in building up fertility. The animal manure is particularly useful to build up fertility on many tropical soils. The effect of manures on crop yield is almost equivalent to artificial fertilizers. Proper conservation and use of these animal manure for a number of years during which the land can be cropped before it is planted into a grass land. No information is available on the length of possible rotations in the tropics. In the tropics, there is a need for maintaining a complete cover. Clean cultivation is the exception rather than the rule in the tropics.

The establishment of grass land will be difficult and need much labour. On the other hand, the eradication of the grass after few years for cultivation is too much for the small farmer, with only hand implements at his disposal. Therefore, it is better that fodder grasses or grasses with legumes are established permanantly for developing grass land in the cultivated land. They could be grown on contour strips as a soil conservation measure. This grasses valuable as a source of cut fodder for feeding green or for conservation as hay or silage.

In Semi-arid and sub-humid Areas

In semi-arid and sub-humid areas, the establishment of perennial grasses may prove difficult. The use of animal manure, combined with cultural methods designed to conserve moisture is a more efficient means of maintaining crop yields than the inclusion of grass or grass/legume pastures. Under these conditions, perennial grasses may seriously deplete the reserves of soil moisture. This may cause reduction in the yield of crops following the grass. Very quick maturing surface rooting, annual grasses are more suitable in these areas. Besides it will protect the soil and control the weeds, thus conserving moisture below the rooting zone. Permanent grass lands may be useful for soil conservation and as a source of fodder for conservation and use in the long dry season.

Tall perennial grasses are useful for fodder as well as for mulching. Where soil moisture is a limiting factor in crop production, mulching is very advantageous. Replacement of annual crops by tree crops or other perennial crops is another approach to the problem of maintaining fertility. The tropical food crops are annual and they cannot entirely solve the problem of maintaining fertility, even in areas suited to their cultivation.

Grasses and Soil Fertility

The inclusion of grass or grass/legume pasture in cropping systems in the heavily populated regions of Asia is likely to be very difficult owing to the need to utilize all the available land for production of food grains for human consumption. Studies m India have indicated that the fertility of soils is generally deteriorating, and even when sufficient fertilizers are used, the yields on many soils cannot be raised to a satisfactory level, further examination has shown that the physical condition of these soils has become very poor, due to continuous grain cultivation with two grain crops per year. Some interesting studies on the introduction of improved rotations in Bihar have been made by the agricultural scientists with the following results.

1. One year grass crop of doob (*Cynodon dactylon*) in combination with annual medics (*Medicago tribuloides*) and *Medicago lupulina* have had a very beneficial effect in improving soil structure and increasing soil nitrogen. In one experiment the yields of the following maize crop was increased nearly three times as compared with the yield following fallow. Even so, the maize yields was only 550 kilogrames per hectare.
2. Doob has proved a much more efficient improver of soil structure than other grasses tried (*Pennisetum purpureum P. Pedicellatum, Chloris gayana, Cenchrus ciliaris* etc.)
3. The beneficial effect on structure of following during the monsoon (rainy season appears to be due to the luxuriant growth of naturally regenerating grasses.
4. The beneficial effect of including on annual fodder grain legume in the rotation depends on the legume grown. *Phaseolus mungo* has greatly increased the yield of succeeding wheat crops while soybean and *Cyamepsis psoraliodes* have been a little or no value.
5. Green manuring for a number of years improved the nitrogen status of the soil but did not increase the yield of the succeeding wheat crop to any considerable extent.

In tropical semi-arid areas, the effect of grasses on soil fertility has received little attention. The black cracking clay soils which are widely distribted in Central India, especially in M.P. and Maharashtra, develop a favourable crumb structure on the surface at the beginning of the rains although these soils are very low in organic matter content even under virgin annual grass land. In this case the structure developed is apparently dependent on the high percentage of montmorillonite clay coupled with a high degree of base saturation and intense drying during the long dry season.

Permanent Grasslands

In many parts of the world the inclusion of grazing crops in the crop rotating is difficult. Because rapid establishment of grass is not possible. The reasons are: low rainfall, high cost of establishment, non-production of seed, difficulty in eradicating the grass in preparation for cropping, necessity for fencing and provision of water supplies, and small fragmented holdings unsuited to grazing.

The ability of a well managed grass strip for controlling erosion and restricting surface run off is well known. Permanent grassland can provide a valuable fodder resource on land unsuitable for cropping. It should be planted on the steeper slopes and in the water ways and gullys where it will serve to protect the cropped land. Narrow strips of fodder grasses planted on the contour between the cultivated fields on sloping land can assist greatly in the control or erosion. This can be border lined with leguminous trees.

Permanent or semi-permanent grasses are included on arable land in the tropics as a source of animal fodder and of bulky organic manures. Tropical fodder grasses in the more humid areas or under irrigation are capable of giving very high yields with only quite moderate amount of fertilizer. Yields of 25 tons of dry matter per hectare are common. An yield from one hectare would provide a maintenance ration for 6-7 mature cattle throughout the year. Crop residues also can be fed along with fodder. This would provide large amounts of organic manure for application to the crop land. Where there is no livestock, grasses are used as source of readily decomposable material for making composts.

The value of animal and other organic manures has, of course, long been recognized by farmers in tropical countries. Many tropical soils have a high power of fixing phosphates, iron and the trace elements. In these circumstances, organic manures may serve a special purpose as a source of reality available plant nutrients.

In India, the value of animal manures, when water is not a limiting factor in crop production, has been proved by numerous experiments. In general, as a source of nitrogen, organic manures have been less effective than equivalent dressings of nitrogenous fertilizers, but in a long term experiment at Coimbatore, farmyard manure which was found to be less effective than completely inorganic fertilizer.

Under dry farming conditions in semi-arid climates, the use of animal manures do no increase crop production. It may in fact reduce yields by causing excessive vegetative growth or by using up the limited soil moisture during decomposition.

Grass Mulch

In the tropics, tall, coarse perennial grasses can be used as a mulch for tropical orchard and perennial crops, particularly shallow rooted tree crops such as coffee and tea. Grass mulch should be laid down before the rains to improve penetration of rainfall rather than after the rains, to retard evaporation as is the usual practice. The yield of clean cultivated orchards could thus be doubled. The rapid improvement of soil structure under the mulch was measured and the reception and percolation of rainfall was improved, while the nutrients applied in the mulch considerably exceeded those removed annually by the fruit and perennial crops.

The problems involved in the application of the mulch are considerable, since the quantities applied were equivalent to approximately 25 tons of sun dried material per hectare; the application of mulch on a large scale would require mechanical cultivation harvesting and transport of the grass crop, and a considerable amount of labour for the actual application to the fruit crops. The labour involved in cultivation is however, reduced owing to the suppression of weeds by the mulch. In a high-value crop such as coffee, spraying of the mulches with weed-killing emulsions containing pentachlorophenol appears to be promising way of reducing the weed. Other tropical perennial crops which have benefited from mulching include Cacoa, sisal, sugarcane (grass, bagasse, trash and cane crops) and rubber, where a mulch is reported to be of particular value in regenerating deteriorated gardens.

The effect of a mulch on annual crops is apparently more variable. As a results of experiment conducted in three different localities, with a grass mulch equivalent to 12.5 tons of dry matter per hectare in forest sites, when not too wet, the mulch had a beneficial effect on the yields of maize and cassava. The effect of the mulch on the groundnut was in general smaller, although crop gave a consistent positive response averaging per cent. In reduced resting period between cropping sequences, the mulch had a greater effect on cereals, a lesser effect on groundnut.

As a conclusion, it may be said that grass mulch is beneficial to the soil and it not only increases the soil fertility but also improves soil physical conditions. In tropical and subtropical areas grass mulches will be good means for sustainable agricultural development.

Conclusion

Grasses are cultivated for fodder/ forage and soil conservation. For forage grasses are grown in pastures. Grasses are cut for green feeding or for storing as hay or silage. Pastures are also used for free grazing. Grasses are cultivated for soil conservation and pastures are perennial types and low yielding. They are maintained usually on non-cultivable land. Forage grasses like napier grass, para grass, sudan grass and bajra are cultivated like crops in good land. All grasses exhaust soil fertility heavily, especially the nitrogen.

Therefore, replenishing nitrogen and other soil nutrients is very essential to maintain productivity of the soil whether in the pasture or in the cultivated lands.

Supplying nutrients with chemical fertilizers periodically and incorporating crop residues and farmyard manure generated by feeding forage and incorporating legume crops are the three main ways of maintaining the soil fertility and grasses.

An indirect method of maintaining soil fertility is never to bum any of the things produced from the land but put back in the form of mulching, animal manure and crop waste. Also protect the grass lands from wild fire. Composting all the crop and grass residues~ all other biomass (like weeds, prunings, trimmings, etc.) and all the farm yard manure before applying into the field is more beneficial.

Incorporating regularly spaced leguminous trees in the pastures and grass lands is a simple but effective means of soil fertility maintenance. Trim all the branches up to 20 ft and allow profuse branching above this height so that grasses at the ground can grow well. On sloping land plant the trees in contour line at 20-25 ft. distance and the levelled land plant them in rows in east-west direction.

In addition to these agronomic practices all the possible soil and water conservation methods can enhance soil fertility. In slopes whatever soil and water conservation techniques are possible should be employed especially in dry and low rainfall areas. It is advantageous to store run-off water or rain water into leaf proof (bottom cemented) tanks. In pasture lands it will serve as drinking water for grazing animals. Otherwise it can be used for irrigating the grasses during dry season.

6

Climatic Requirements of Plants

Climate is one of the main natural factors governing the plant growth and development. It is also a primary factor which determines , tile success or failure of the plant life. Climate can be defined as the sum total of all atmospheric or meteorological influences, Ii principally temperature, moisture, wind, pressure and evaporation which combine to characterize a region and give it individuality by influencing, the nature of its land forms, soils, vegetation and to the land use.

Climate has a great role to play in almost all the agricultural different activities from the initial preparatory tillage to the harvesting and storage of produce. Even the price trends of agricultural products life. are also controlled a lot by the climate. Climate and weather are the two major limiting factors (in crop production) to be considered in the matter of the time of sowing, transplanting, scheduling of irrigation, timing of fertilizer application, using of pesticides etc. Weather is the day to day condition of the atmospheric temperature and the presence of rain, sunshine and wind. Thus weather of a booklet; place is said to be dry, cold, hot, warm, sunny, wet, cloudy, clear, windy etc. whereas climate is the, regular pattern of weather, conditions extending over a period of time in a particular region. The climate of a particular region is fairly uniform over a wide variety and area. However, variation in the climate can take place in relatively small areas mainly due to the topographical or other factors. These local variations which are limited to small areas and create a small climate are called micro-climate.

A sound knowledge or the climatic factors and understanding of the complex processes of interaction between climate and biological processes of the plant are essential for a scientific approach to farming The following are the climatic factors that affect the crop plants in their growth and development.

Light (Solar Radiation)

Lightis the most important abiotic (non-Iivmg) factor without no Iife can exist. Climatic factor also accelerates many vital processes of the plant. The chief natural sources of light are sunlight, moon light and star light. Of these sunlight has the greatest ecological significance. In fact, farming is simply the harvesting of solar energy through crop plants.

Solar energy meets two essential needs of plants: (I) Light required for photosynthesis; and (2) thermal conditions required for the normal physiological functions of the plant. These are needed for the functions such as seed germination, leaf expansion, growth of stem and root, flowering, fruiting, and reduces the dormancy. Light is an indispensable factor for the synthesis of increases the carbohydrate in plant cells. This sunlight is absorbed by the green, pigment of the plant i.e. chlorophyll. Chlorophyll captures the solar energy from sunlight and preserves it in the form of carbohydrates which are the chemical compounds produced inside the leaves by the process called photosynthesis. The carbohydrate manufactured by by the plants during photosynthesis is the only link between the solar energy and the living world.

The whole photosynthesis process can be summarized in the following formula:

$$6\,CO_2 \;+\; 12\,H_2O \xrightarrow[\text{Chlorophyll}]{\text{Solar energy (673 cal)}} C_6H_{12}O_6 \;+\; 6O_2 \;+\; 6H_20$$

Carbon + water Carbohydrate + Oxygen + water dioxide or glucose

Photosynthesis consists in the building up of simple carbohydrates such as glucose and fructose in the green leaf by the chloroplasts in the presence of sunlight by using carbon dioxide and water as raw material. All the four characteristics of light namely light intensity, light quality, light duration and light direction influence plant growth and development These are discussed below:

Light Intensity

The variation of light intensity takes place when there is a the change in temperature and relative humidity, which controls the process of photosynthesis very much. The very low light intensity reduces the rate of photosynthesis and may even result in the closing of stomata. This will retard the growth of the plant. Similarly, very high light intensity is also detrimental to plant in many ways. It increases the rate of respiration and this disturbs the photosynthesis-respiration balance. It causes rapid loss of water from plant leaves resulting in the closure of stomata thereby blocking the entry of CO2 into the leaves. The high light intensity also causes the by phenomenon of solarization in which the cell contents are oxidised by atmospheric oxygen resulting in the loss of CO2 from plant the leaves.

Most of the crops show marked response to change of light the intensities. During cloudy weather the production of dry matter is limited. It also depends on the type of plants. Some plants are heliophytes (sun loving) and some are sciophytes (shade loving). Most of the species produce maximum dry matter under high light intensity if water is available in plenty. However, crops like betel, sunflower, wheat and tobacco, grown during summer, produce greater dry matter if slightly shaded.

Quality of Light

When light is passed through a prism it is dispersed into wave lengths of different colours viz., violet, indigo, blue, green, yellow, orange and red. The most effective wave length for the growth and development of a plant is red light followed by violet and blue. Red light is especially important for the photosynthesis.

Many factors such as amount and kind of cloud cover, fog, hence they air pollution and the colour of the foliage, influence the light and its impact on plant growth and development.

Duration of Light

Duration of light is particularly significant in the selection of crops and their varieties for a particular region. Duration of light influences the flowering, fruiting, bulb formation and rhizome production in plants. Different crops show different responses to the relative length of day and night and this is termed as photoperiodism. Depending on the requirements of day length, crop, can be divided into the following three categories:

Short-day Plants

Flowering in these plants does not take place above the critical photoperiod. These plants require longer nights and shorter days for their flowering. For instance, if critical photoperiod, (the minimum amount of daily light required by certain plants for their growth and flowering) (day length) is 8 hours for the flowering in these plants, it should not be beyond this limit, or else vegetative growth will continue and flowering will remain inhibited. However, flowering may occur below the critical photoperiod.

Long-day Plants

These plants require longer days and shorter nights for their proer flowering. These plants will not start flowering below, minimum critical day period.

Day-neutral Plants

Flowering in these plants is not influenced by the day length, hence they are called day neutral plants. Flowering in these plants may occur both in long-day conditions as well as short-day conditions.

According to their relative photoperiod requirement, some crop plants are given in the following table.

Table: *Photo periodicity or different crops light*

Short day plants	*Long day plants*	*Day neutral plants*
Maize,soybean, jowar, bajra, tobacco, lobia, green gram (moong) etc. strawberry, etc.	Wheat, barley, oat berseem, pea, gram cowpea, sugarbeet, etc.	Couon, sunflower, tomato, asparagus,

However, paddy is basically a short-day plant but some dwarf varieties like IR-8, Jaya, Padma etc. are photo-insensitive varieties.

Direction of Light

Shoots, roots and leaves show different tendencies to the direction of light. In temperate regions, it is clearly evident from the fact that the crop growth is better in southern slope than northern because the former receives more sunlight. This is especially more important for the crops which are highly photo- sensitive in nature.

Importance of Light to Plant

Lights affects many physiological activities of the plants such as photosynthesis, respiration, opening and closing of stomata, growing and flowering of plants, movements and germination of seeds.

Photosynthesis

Light plays an important role in the development of chloroplast (containing green pigment chlorophyll) in the leaves. Their number and position in the leaf is also largely determined by the availability of light. According to the direction of sunlight the leaves arrange the chloroplast in line. At high light intensity, the photo-oxidation of enzymes may reduce their availability for photosynthesis and (protein synthesis, thereby badly affecting these vital processes.

Respiration

There is no direct effect of light on the respiratory activities in the plant body. But photosynthesis, which occurs only in the source presence of sunlight, has an influence on respiration. Under certain conditions, for example, in shade, the light bec6mes a limiting factor and the rate of photosynthesis may or may not be sufficient to meet the need of respiration as the oxygen which is produced during the process of photosynthesis is used to decompose the complex organic substances like carbohydrates, fats and proteins. 'germinate Therefore, in some plants, respiratory rate is minimized by removing the plant leaves after fruiting has taken place. The bright sunlight also keeps stomata open and oxygen is easily and quickly absorbed.

Opening and Closing of Stomata and Transpiration

Mostly the stomata remain opened in the light and closed in the dark. Light brings about conversion of starch into soluble sugars in the guard cells and thereby increases their osmotic pressure which in turn causes flow of water in the guard cells. The increase in the turgidity of guard cells causes widening of gap between two guard cell. The opening of stomata increases the gaseous exchange as also the rate of transpiration.

Growing and Flowering of Plants

High intensity of light inhibits the production of auxins and other growth hormones. Consequently, growth and development of plants is seriously affected. On the other hand, plants growing in darkness or insufficient light produce maximum amount of growth hormones as a result of which they grow tall and lanky with slender pale yellow stems and small leaves.

Movement

The stems, shoots, roots and other parts of the plant show different response to light The leaves and stems grow towards the source of light.

Germination of Seeds

The moistened seeds are very sensitive to light. In some cases the germination of seed is retarded in light. In fact, seeds germinate more quickly in the dark. Seeds of tomato will not germinate unless they are kept in the dark. However, in the case of tobacco, seeds should be treated with sunlight sowing.

Precipitation

Fresh water supply to the earth is almost entirely met as precipitation resulting from evaporation of sea water and land. In the an surface. It occurs in a variety of forms such as rainfall, snow, hail, dew, fog etc.

Precipitation affects the vital process of all the living beings. It is the plenary agent sets in motion the nutrients of the soil and effect or makes them available to plants. It affects the morphology and physiology of the plants as well.

The total amount and seasonal precipitation greatly determines choice of cultivated crops of a place. Dry fanning is practiced in areas which low and poorly distributed rainfall occur. Only moisture drought resistant crops (sorghum, pearl millet, barley etc.) are grown in such regions. On the other hand in the places of heavy and regular rainfall areas, crops such as rice, tea, coffee, rubber etc. or later are grown. However, heavy rainfall can also be disadvantageous as it increases the leaching of nutrients or carries away soil and applied fertilizers along with surface run off. Cool season rainfall is more beneficial to crops than the crops grown during the warm season, when evapo-transpiration is at its maximum. Similarly, light and prolonged rainfall is more beneficial to plants than the quick and heavy ones. If the ground is frozen, infiltration will be low or nil. If the crop canopy is dense and broad leaved (vegetation is in full leaf), a larger portion of the rain falling down will be (intercepted and lost by evaporation before reaching the soil. There are various forms of precipitation which affect the plant growth and development. (rainfall is the most important form of precipitation in the regions with tropical and sub-tropical climate as in India. By and large the distribution of rainfall determines the quality of agriculture in a particular place.

Rainfall

Variability of rainfall is the greatest hazard to crop production in the arid and semi arid regions. In good rainfall season cultivators may plough and sow large area of their land, but if it is followed by scanty rainfall season with increase in temperature, it will prove disastrous to crop production. On the other hand, very light rainfall will only wet soil surface which will not have any appreciable effect on the soil moisture requirement. Therefore, the rain must penetrate into the soil to a depth of at least 10-12 cm. The amount needed for this has been estimated to be 15-20 mm for a single rainfall.

The distribution of the rainfall and adequate supply of soil moisture at periods of maximum crop requirement also have bearing on yield levels. Drought at the time of earing of cereals will have more adverse effect on yields then if it occurs earlier or later during the growing period. Excess moisture after sowing may reduce germination. Heavy rainfall during earing may cause lodging and impair pollination.

Rainfall is also related to other agro-climatic condition such as the degree of evaporation and transpiration, intensity of soil moisture deficit and characteristics of soil storage and replacement.

Snow

Snow has its own bearing on livestock and cropping. There Snowy winters on higher latitude and altitude cause a severe loss to pastures and crop field by depositing thick layer of snow on ground. It also affects the hill sheep farming. The frozen ground may cause adverse condition for cropping because certain parasitic fungi find favourable condition for breeding under snow cover and attack plant seed in the soil, when they are grown. These pathogens may also be transported by air to snow-free places and cause plant diseases such as late blight of potato.

Dew and Frost

Dew and frost have three different effects on plant growth. First they reach the soil when water runs off the leaves, and thus make themselves available to plants. Secondly, water is absorbed by the leaf surface on which it condenses and is used directly by indirectly the plants. Thirdly, they boost the growth of plants by delaying the rise in leaf temperature on the following morning, thereby acting down the rate of evapo-transpiration.

Fog and Mist

Fog is a very thick mist formed either by the evaporation of warm water in sufficiently cold air, or cooling of air while in contact with the ground. Mist is caused by the condensation of water vapour in the air.

Non-irrigated winter crops positively benefit from the amount of moisture supplied by fog, as it is much greater than dew. It affects plant growth through the high air humidity, through wetting of the aerial parts of plant, and through humidification of the surface, of the soil.

Temperature

Temperature is a variable factor and is influenced by time, season, latitude, altitude, slope direction, soil texture, plant cover and human activities like urbanization and industrialization. It penetrates every region of the biosphere and exerts its impact by increasing or decreasing some of the vital activities of plant, such as metabolism, behaviour, reproduction, embryonic development and growth.

Each crop plant needs a certain amount of effective heat units for germination, growth, stalking, maturity and ripening. This is called thermal constant and it varies from crop to crop. The crucial air temperature is 6 degree centigrade below which the plant does not grow. Ideal temperature conditions for most crops is between 18.3 and 23.9 degree centigrade.

Table: *Ideal temperature for the crops*

Sl.No.	***Crop***	***Suitable temperaturerange***	***Rainfall***
1	Paddy	20-35	100
2	Jowar	25-35	40-60
3	Jajara	25-35	35-50
4	Maize	30-40	40-60
5	Wheat	16-25	40-150
6	Barley	16-25	20-100
7	Gram	15-25	40-80
8	Arhar	25-35	75-125
9	Groundnut	22-30	80-120
10	Sesamum	20-30	40-60
11	Mustard	16-25	20-60
12	Linseed	16-25	50-75
13	Coltton	32-40	75-115
14	Sugarcane	20-30	100-150
15	Potato	15-25	50-100
16	Tobacco	15-30	40-60
17	Jute	25-35	100
18	Tea	20-30	150-200
19	cocoa	15-30	200

The temperature affects the vegetation either directly or indirectly. Directly it affects in two ways.

1. It regulates the physiological processes of plants and consequently their growth and size.
2. It determine which species can grow in a particular region.

The temperature above 40° C and below 6° C is extremely harmful for plant growth and development. In these extreme ranges of temperature, different kinds of adverse soil conditions and plant injuries occur, sometimes killing the plant as such. The following are some of the examples.

Cold Injury

Chilling Injury: Plants growing in hot climate if exposed to low temperature (above freezing point) for some time, are found to be killed or injured severely.

Freezing Injury: When the temperature falls below the freezing point, the water will be transformed into ice crystals in the inter-cellular spaces of the plants. In such a condition, the protoplasm of the cell is dehydrated and mechanical distortion of the tissues takes place if temperature falls further. This may ultimately result in the death of the plant

Suffocation: During winter the ice or snow form a thick layer and cover the ground, and the crops suffer for want of oxygen.

Heat Injury: When the temperature is very high, plant growth is stopped and respiration rate becomes high. In such a condition, plant may die due to starvation. The effects of high temperature are more pronounced in young seedlings and transplanted crops. Besides, hot-dry winds cause a severe damage to the flowering and fruiting crops. Young plants are adversely affected by these desiccating winds.

Temperature in combination with humidity and other factors helps in the spread of diseases in plants. Low temperature, high humidity and cloudy weather favour the damping off, seedling blight, foot rot and root rot diseases of cucurbits, tobacco, papaya and ginger.

With increasing temperature, evapo-transpiration increases. Hence a moisture supply that is adequate in a cooler climate may not suffice in a warmer region.

Atmospheric Humidity

Water is always present in the atmosphere in the form of vapour. This is known as absolute humidity of the air. The amount of water required to saturate the same unit volume of air under constant physical conditions is called relative humidity. This is expressed in terms of percentage. The amount of water vapour in the air is highly variable depending on the existing temperature and also on the nature of its source which may be ocean, desert, grass land and forest.

The atmospheric humidity influences directly the form and structure of the plants. It directly affects the transpiration rate of the plants. In dry atmosphere transpiration rate increases and as a result. the water content of leaf tissues decreases and the leaves wilt temporarily.

Atmospheric humidity is most significant in two ways. First, plants can absorb moisture directly from an unsaturated air of high humidity. Secondly, humidity affects the photosynthesis in the leaves of plants. Rate of flowering in certain plants (eg. pea nut) increases with increase in humidity. Some plants gain more weight and well developed root system under high humidity condition. (Eg. maize and tomato). The efficiency of water use would increase with more humidity in the air. Moist air favours the growth of many fungi plants are and bacteria and thus affects seriously the crop. The blight diseases of potato and tea are common examples of diseases spread under moist weather. Many kinds of insect parasites such as aphid and jassids thrive well under moist condition.

Wind: Air may travel a great distance depending on their force and track. Wind has many direct and indirect influences on crops and plants.

Following are the important effects of wind on vegetation.

1. Wind causes erosion of soil and abrasion of vegetation through removal of soil particles, and decreases the growth of plants by way of reducing the moisture content of air.
2. In strong dry and hot wind, young plants may shrivel and roots of young plants may get exposed, sometimes uprooting the whole plant.
3. Wind may cause considerable loss by inducing lodging, and breaking of stalks and shattering of grain.
4. Physiological effects of wind consists mainly in increasing as transpiration as well as evaporation from the soil.
5. Wind is an important agent-for the dispersal or pollen grains, fruits and spores of the plants.
6. In desert storm results in big sand dunes which cover the vegetation causing further desertification.

Since wind velocity increases with height above the ground, the leaves near the soil surface remain comparatively safe from the drying effect of the wind. On the other hand, the tall horticultural plants are the worst hit from excessive transpiration. In India the monsoon type of rainfall is largely determined by particular patterns of wind movement

Atmospheric Gases

The earth is surrounded by a gaseous layer called atmosphere. Gaseous mantle atmosphere extends into outer space some 1000 km or so above the earth surface.

The atmosphere surrounding the earth contains a mixture of gases. The composition of this atmosphere with the main gases Ion. are given as follows.

Table: *Constituent of Air Percentage by volume*

Sl.No.	*constituent*	*% by volume*
1	Nitrogen	78
2	Oxygen	20
3	Carbon dioxide	0.03
4	Others	1

Nitrogen

Nitrogen constitutes the major part of the atmospheric gases. The nitrogen is made available to the soil by lightning, rainfall respiration and nitrogen fixing micro-organisms. Free living bacteria like above 70% azotobacter, blue green algae fix a good amount of nitrogen in the cases of soil. The decomposition of the protein content of dead bodies of plant and animal by micro-organisms is also a good source of nitrogen to the soil.

Carbon dioxide (CO_2): Along with water, it is the main raw material for synthesis of carbohydrates through the process of photosynthesis by the green leaves of the plants. Therefore, its proper concentration in the atmosphere is of paramount Importance. Up to a certain limit, photosynthesis increases with the increase in CO2 content.

Increased growth and greater yield of vegetables are possible under glass house conditions by increasing the CO_2 concentration in the soil may cause suffocation of the roots resulting in poor root development that will ultimately affect the plant.

Oxygen (O°2): Every living organism whether plant or animal sustains its life because of oxygen. Atmospheric oxygen is vital to the life and normal growth of plants. The amount of oxygen is normally constant in the air because plants release oxygen during photosynthesis.

Oxygen is necessary for the formation of chlorophyll. Seedlings fail to develop chlorophyll in the absence of oxygen even when these are exposed to sunlight. In plants, oxygen is necessary for respiration, a process that competes with photosynthesis. Higher the oxygen availability, greater the respiration rate and lower the photosynthesis. Oxygen concentration above 70% stops photosynthesis completely. However, extreme cases of high or low concentration of oxygen normally do not occur nature.

Conclusion

The success or the failure of the crop production is determined is by the intensity of the climatic factors. The climatic conditions are the primary factors controlling the distribution of crop and agricultural patterns in a particular geographical area. So there always exists a significant relationship between climate and crops.

Therefore, for obtaining meaningful results there is a need for collecting enough information about the visible and invisible m climatic elements and their impact on agriculture in the regional perspective. Such information should be recorded and made available to the farmers in the form of daily, weekly, fortnightly, monthly, seasonal and yearly handouts. If the farmers are informed about the possible climatic development well in advance, they can make an accurate plan of farming activities for a specific season and locality. If they are given prior information about the onset of monsoon, they will be able to resume their farming activities in right time and may also escape the vagaries of atmospheric fluctuation. And the ultimate and most important objective of increasing farm productivity or the income of poor farmer can be realized only when all these factors are no longer problems to our farmers.

Integrated Nutrient Management

Integrated nutrient management (INM) is an integral part of the sustainable agriculture which requires the management of resources in a way to fulfill the changing human needs without deteriorating the quality of environment and conserving vital natural resources. By the turn of this century, the targetted crop production is about 225 million tonnes to feed over one billion population of our country. So this has to be achieved without deteriorating the land by mis-management. At the same time we have to limit our population. However, it will take decades before we can bring down the rate of growth of population. When the crop growth and maturation is over and the crop is harvested that much nutrients are removed from the soil. Hence there is the need for replenishing the soil by supplying nutrients as there would be nutrient utilization during the next crop development. This can be supplied by chemical fertilizers. The strategy for sustaining satisfactory yield levels, envisages nutrient balances and efficient nutrient cycling. Every plant requires nutrition which may be supplied in the form of fertilizers, organic manures, bio-fertilizers, green manures, etc.

The use of high yielding crop varieties and chemical fertilizers has resulted in rapid increase in agricultural production, especially in irrigated areas. But complete reliance on the use of fertilizers ignoring bio-organic materials affect the soil environment. Indiscriminate use of chemical fertilizer will lead to:

1. Widespread nutrient deficiencies in soils,
2. Disturbs soil reaction,
3. Development of nutrient imbalances in plants,
4. Increased susceptibility to pests and diseases,
5. Reduces the soil organic matter,
6. Harmful to beneficial soil micro organisms,
7. Reduces legume root modulation and plant mycorrhizal association
8. Increases environmental pollution.

The realization of these effects of continuous use of fertilizers in large quantities in the absence of organic supplements has triggered a new interest in the Integrated Nutrient Supply System or Integrated Plant Nutrient System (IPNS)

Integrated nutrient management comprises of application of organic manures, green manures, blue green algae, bio-fertilizers and crop rotation with legumes along with minimum use of chemical fertilizers to produce optimum crop yields without deteriorating the soil health.

Integrated use of fertilizers and organic materials add organic matter to the soil which is essential for the healthy development of soil micro-organisms. They have the ability to carry out biochemical transformations and serve as a source and sink for mineral nutrients. They have an active role in decomposing organic matter and helps in releasing the essential plant nutrients. Besides this, organic matter is very important in several ways,

a. If plays a vital role in maintaining soil fertility and keeping the soil productive.
b. It improves soil structure.
c. It increases the microbial activities.
d. It increases water holding capacity of the soil
e. It supplies macro and micro nutrients.
f. It enhances the nutrient availability.
g. It increases the water use efficiency of crops.
h. It improves the aggregation of soil particles.

Integrated nutrient management necessarily implies knowledge about organic manure (Booklet No. 31), Farmyard manure (Booklet No. 63), Composting (Booklet No. 64), Green manuring (Booklet No. 66), Introduction to fertilizers (Book- let No. 15), Fertilizer recommendations (Booklet No.19), Application of fertilizers (Booklet No. 23), Fertilizer use in Dry- land Agriculture (Booklet No. 43), Biofertilizers (Booklet No.3).

Components of Integrated Nutrient System

Nutrient can be supplied through fertilizers and manures. But manures contain very less percentage of major macro nutrients such as nitrogen, phosphorus and potassium com- pared to fertilizers. They will not be able to meet the nutrient requirement of the crops, when applied alone. So we cannot totally depend on the manure and they should be integrated with the commercial fertilizers for fulfilling the needed quantities by plants. This integrated approach not only meet the requirement but also adds organic matter to the soil which acts as a buffer in maintaining the soil environment and thus the healthy growth of the plants. The major components that can be used in the integrated nutrient management are: organic manures, legume planting, bio-fertilizers and others residues and biproducts.

Organic Manures

Organic manures are available in the form of green and dry plant residues fresh animal wastes, decomposed materials of plant and animal origin and biologically active preparations. When added to soil they undergo microbial decomposition. In this process, the nutrients held in organic combinations are slowly released in available forms besides improving the availability of nutrient elements present in the soil. They also promote the microbial and soil enzyme activities.

Farm Yard Manure

This manure is produced in the farm mainly with animal excreta. It is made up of the excreta of farm animals the litter or bedding provided for them and miscellaneous farm and house hold wastes. Cattle dung, excreta of other animals like sheep, horses, goats, poultry, etc. are collected and can be utilized as a manure. Litter is a bedding material that is spread in animal houses and the urine get absorbed into the bedding material. Straw, sawdust, peat, dry leaves etc. are used as bedding materials.

Cattle are the main livestock maintained by the farmers in the country and the manure produced in farms may appropriately be called as cattle manure. It can be collected and can be made into manure by three main systems, namely dry earth system, byre and pit system and loose box system.

In dry earth system, dry loose loamy soil of 15 cms thickness is spread on the floor of the cattle shed. The urine soaked earth along with dung is removed everyday and stored in the manure pit.

In byre and pit system, cement or stone floor of the cattle shed is gently sloped towards manure pit by channels that will lead the urine into the pit.

In the loose box system separate pits are made and spread with a layer of straw. Animals are tethered there for dunging. Dung is not removed daily but spread on the straw. The urine is absorbed by the straw. Another layer of straw can be spread and let the animal dung and urinate on it. This process can be continued till the pit is full (may be in six months) and decomposed material can be directly taken to the field for use.

Composts

All organic farm wastes, stubbles of crops waste straw, sweepings, threshing floor collection, dry weeds etc. may all be collected and used for augumenting the organic matter supplies they cannot be used directly as manure in most cases. They require to be decomposed before application. The process of decomposing wastes is called composting and the decomposed material is called 'compost'. It is like well decomposed cattle manure, more powdery and lighter; in colour.

Compost making involves keeping the dry organic waste materials in layers in pits and moistening either with water or with cow dung slurry. It takes about 8-10 weeks for complete decomposition. There are several methods which can be seen in Booklet No. 64. Besides these methods there is another method which was developed by P AE department. For details refer booklet No. 146.

Green Manures

Growing a crop purposely and incorporating it in the soil for manuring is called green manuring. Collecting green leave from all available sources and using it for manuring is green leaf manuring, Legumes alone are used as green manure crops usually. Ex. *Dhaincha, Sesbania speciosa, S. rostrata,* wild Indigo (*Tephrosia purpurea*), indigo (*Indigofera tinctoria*), *Phaseolus trilobus,* sunhemp, etc.

Concentrated Organic Manures

Concentrated organic manures are the organic manures which are rich in nutrients than bulky organic manures. Ex. guanos, fish manures, bone meal, oil cakes, etc. can be used in integrated nutrient management.

Legume Planting

Including legumes in crop rotation is a very beneficial practice. Legumes are the plants which have nodules on their roots in which rhizobium (a bacteria) lives. This bacteria have the ability to fix atmospheric nitrogen into the root nodules. They live in symbiosis with the plant Many other free living bacteria like *Azotobacter, Beizerinckia, Clostridium,* etc. and blue green algae like *anabaena, nostoc* etc. are also beneficial.

Bio-fertilizers

Bio-fertilizers are defined as biological preparations containing live or latent cells of efficient strains of nitrogen fixing or phosphorus mobilizing micro organisms. They are also known as microbial inoculants.

Rhizobium

In recent years bio-fertilizers have emerged as an important component of integrated plant nutrition system. The largest contribution of biological nitrogen fixation to agriculture is derived from the symbiosis between legumes and rhizobium species. Inoculation with efficient strains of rhizobium specific to each crop is essential for the nitrogen gains and better crop yields.

Nitrogen fixed by legume-rhizobium symbiosis is 100- 300 kg. nitrogen per hectare.

Blue Green Algae (BGA)

In Indian soils, the predominant genera are *Anabaena, Nostoc, Calothrix*') etc. BGA inoculation add up to 50 t/ha of organic matter and contributes 20-30 kg. nitrogen per hectare.

Azotobacter

Azotobacter is a free living bacteria which not only provides nitrogen but also produces a variety of growth promoting substances like indole acetic acid, gibberillins, B-vitamins and anti-fungal substances. Observations indicated that inoculation by Azotobacter was significant and it sprayed from 34-247 kg. N per hectare indifferent locations.

Azospirillm

Inoculation of *Azospirillum* has shown positive response in several field crops with an average response equivalent to 15.20 kg. nitrogen per hectare.

Mycon-hiza

Vesicular *Arbuscular Mycorrhiza* (VAM) are formed by non-septate fungi belonging to the genera *Glomus, Gigaspora, Entrophospora*, etc. it enhances the phosphorus availability and also the availability of other elements like potassium, sulphur and micronutrients like copper, zinc, aluminium, manganese, iron, etc. Mycorrhizal colonisation also allows introduced population of beneficial soil organisms like *Azotobacter, Azospirillun* and phosphate solubilizing bacteria to thrive. Reports show that VAM fungi can reduce fertilizer requirement by 25-30 per cent.

Phosphatic Bio-fertilizers

Phosphate solubilizing organisms such as Pseudomonas striata, Racillus polymixa, Asperigillus awamori, Pencillium digitatum etc. can grow with insoluble phosphate sources and convert them into soluble forms by a variety of mechanisms, Indian Agricultural Research Institute, New Delhi has prepared carrier based inoculant known as "IRRI Microphos" culture using efficient strains of *P. striata* and *B. polymixa.*

Others

There are several others which can be used in integrated supply system. Easily decomposable crop residues can be incorporated into the soil which add nutrients and organic matter. Crop residues of annual grain legumes such as green gram, Cowpea, black gram, horse gram and soybean can be ploughed back into soil after the grain is removed. Addition of organic residues would be more meaningful in dry lands since it is not possible to sustain a build-up of organic matter in the soil due to rapid decomposition.

Another waste material, in the coir industry, coif pith is a potential source of organic manure in India. Coir pith is rich in potassium and micronutrients such as iron, manganese, zinc and copper.

Pleurotus sojor-caju, Aspergillus and Trichoderma are found to be potent degraders of coir pith. They are used to hasten the conversion of coif pith into compost. Addition of one kg. used and 300 g pleurotus to every 200 kg. coir pith will result in good compost in about 30 days. It is used in both irrigated and drylands areas but more useful in drylands as it has high moisture conservation efficiency. Different industrial wastes like flyash, industrial effluents, etc. urban garbage, saw dust, sewage sludge and biogas slurry can also be used.

Successful Applications

Trials with a number of permutations and combinations with fertilizers and organic manures showed that integrated nutrient management is a success and it contributes towards sub stainable agriculture. As it includes the use of organic materials along with fertilizers, it adds organic matter to the soil which is a major component of the soil system. Continuous use of organic manures slowly and gradually increases the nutrient content in soil thereby reducing the fertilizer dos- age.

Soil should be considered as an organic mass and the organic matter is the soul of soil. Just as a body without soul is dead so also soil without organic matter is a dead material meaning to say that it is an unproductive material. Besides the negligible amount of NPK nutrients organic matter provides the most valuable and essential micro nutrients which the chemical fertilizers are unable to provide. At the same time the organic fertilizers are unable to provide all the required quantity of NPK nutrients. Therefore both organic manures and chemical fertilizers are used in integrated nutrient management. A minimum of 10-12 tonnes of organic manures should be applied per hectare.

Farm Trials in Haryana

In paddy, experimental plots were treated with green manure and NPK (nitrogen, phosphorus and potassium). legumes pulses are normally used for green manuring. Different combinations of NPK fertilizers and combinations with green manure was tried. Green manuring with Dhaincha (*Sesbania aculeata*) was incorporated at the rate of 12.5 tonnes per hectare before transplanting of paddy the yields of various treatments.

Table: *Yield of paddy (t/ha) in farm trials in Haryana*

Sl.No	*Districts*	*Green manure+NPK 120:30:30*	*NPK 180:30:30*	*NPK 120:60:60*
1	Karnal	7.00	6.99	6.58
2	Kurushetra	7.18	7.10	6.79
3	Sonepat	6.54	6.47	6.19

This indicates that green manuring with NPK has not only given higher yield but also reduced the nitrogen fertilizer application from 180 kgs. to 120 kgs. Though fertilizer application alone can give higher yields, the fertility and productivity can be maintained only when sufficient quantity of organic matter is present in the soil.

In this trial, succeeding wheat crop also yielded higher on green manured plots, confirming the residual effect of green manuring.

Integrated Nutrient Management with bio-fertilizers: Bio-fertilizers are of biological origin and serves the function of supplying the nutrients for effective plant growth. Several trials are given here.

Paddy Yield due to BGA Inoculation

The various studies conducted in Indira gandhi Krishi Vishwa Vidyalaya, Raipur the efficiency of BGA (Blue Green algae) was found decreasing with increasing dose of nitrogen and maximum increase in yield was recorded in combination with 20 kg. N per hectare.

Large scale on farm trials conducted in Tamil Nadu and Madhya Pradesh had indicated increase in yield when BGA was used. Table explains the same.

Tahle: *Average paddy yield (t/ha) due to BGA (10 kg./ha)*

Sl.No	*Treatment*	*Madhya Pradesh*		*Tamil Nadu*	
		Gain	*% increase*	*Gain*	*% increase*
1	Zero N:40P:15K	24.1			
2	Zero N:40P:15K + BGA	28.2	17.0		
1	100 N:50P:50K			51.1	
2	100 N:50P:50K + BGA			54.8	7.2

Effect of bio-fertilizers on Yield of Sugarcane

A detailed study conducted by Sugarcane breeding Institute, Coimbatore on different locations and soil types has shown that inoculation of *Azospirillum* with 180 kg. N per hectare could give an yield of 106.4 t/ha which compared well with what was obtained at the level at 240 kg./ha of inorganic nitrogen. It indicates a saving of 60 kg. nitrogen per hectare. Azospirillum also perfom1ed well in improving the sucrose percent in juice.

Table: *Response of sugarcane to azospirillum and azotobacter inoculation*

Sl.No	*Treatment*	*Cane yield t/ha*	*Sucrose % in juice*
1	120 kg N/ha + Azospirillum	93.2	16.12
2	120 kg. N/ha + Azotobacter	80.8	16.22
3	180 kg. N/ha + Azospirillum	106.4	16.4
4	180 kg. N/ha + Azotobacter	93.8	16.13
5	240 kg. N/ha (urea)	99.6	16.24

In another experiment under water stress condition, effect of soil and set application of *Azotobacter* and *Azospirillum* were studied over fertilizer application alone. *Azotobacter* was found to be a better performer than *Azospirillum*. The yield due to *Azotobacter* along with 225 kg. N/ha was found equal to that of 300 kg. N kg/ha alone, thus saved 75 kg N/ha.

Azotobacter and *Azospirillum* have been applied in two split doses to soil at 30 and 60 days after planting.

Table: *Effect of biofertilizers on yield and quality of surgarcane*

Sl.No	*Treatment*	*Cane yield (t/ha)*	*Sugar (t/ha)*	*Net returns Rs/ha*
1	300 kg. N/ha (recommended dose)	92.5	11.52	23,075
2	Azoosdpirillum + 75% N	87.81	10.85	21,298
3	Azospirillum + 50% N	77.26	9.63	17,304
4	Azotobacter + 75% N	94.6	11.94	23,947
5	Azotobacter + 50% N	89.07	11.36	21,916

Integrated Nutrient Management with Manures: There are a number of experiments conducted on this, some of which are given here.

Permanent Manurial Trials

Permanent manurial experiments were started because they give sufficient time to study phosphorus, potassium and other secondary and micro nutrient needs of crops. Seasonal experiments are helpful to study mainly the response of nitrogen fertilizer. So, in order to study the effect of continuous application of fertilizers on soil health permanent manurial experiment is the only way. Since 1909, a permanent manurial experiment is being carried in Coimbatore Agricultural CoIlege. The treatments taken are as follows. The cropping system includes cotton, ragi, sorghum and dewgram.

1. Control (no manure)
2. Nitrogen (N)
3. Nitrogen and potassium (NK) (Recommended doses)
4. Nitrogen and phosphorus (NP) (Recommended doses)
5. NPK(Recommended doses)
6. KP (Recommended doses)
7. Kalone (Recommended loses)
8. P alone (Recommended doses)
9. Cattle manure (12.5 t/ha)
10. Cattle manure residue

Observations indicated that cattle manure treatment was on par with NPK treatment in increasing the yields of cotton sorghum and dew gram.

Another experiment was conducted with a fixed cropping sequence of ragi-maize- cow pea with 10 treatments.

1. 50% optimum NPK
2. 100% optimum NPK
3. 150% optimum NPK
4. 100% optimum NP
5. 100% optim1.UI1 NPK + hand weeding
6. 100% optimum N alone
7. 100% optimum NPK + FYM 10 t/ha
8. 100% optimum NPK
9. 100% NP + Zinc sulphate 25 kg./ha
10. Control (no fertilizer and no manure)

The yield of ragi, maize and cowpea crops was studied in the 14th cropping cycle. The combined treatment with NPK + FYM recorded maximum grain yield for all the crops. Continued application of nitrogenous fertilizer alone significantly reduced the yield because of phosphorus which became a limiting nutrient. When nitrogen was applied in combination with phosphorus yield increase was observed. When NPK fertilizer dose was increased from 50% to 100% there was appreciable increase in yield but for further increase to 150% there was not much additional yield.

The analysis of post harvest soil at the end of fourteenth cropping sequence indicated that when phosphorus was not added continuously depletion of native soil phosphorus was observed limiting the crop growth. The combined application of NPK + FYM recorded higher available macro and micro nutrient status.

Integrated Management with Farmyard Manure

In soybean, an experiment was conducted to study various characteristics. To one plot recommended doses of NPK (20-40-40 kg./ ha and the other treatment was farm yard manure with half of the recommended dose. (FYM + 10-20-20 kg/ha). Nitrogen was applied in split doses of three.

Table: *Integrated nutrient management in soybean*

Sl. No	*Parameters*	*Nitrogen 20/40-40 kg/ha*	*FYM + Nitrogen 10/20-20 kg/ha*
1	Seed yield (kg/ha)	1697	2044 (+ 20.5%)
2	Water use efficiency (kg/ha mm)	3.71	4.5 (+21.3%)
3	Nitrogen (kg/ha)	124	152 (+22.6%)
4	Phosphorus (kg/ha)	11	13 (+18.2%)
5	Potassium (kg/ha)	42	68 (+61.9%)
6	Sulphur (kg/ha)	10	14 (+40%)

Figures in parentheses indicate the percent increase over fertilizer treatment. Experiment also proved that when farm yard manure and blue green algae were applied along with the fertilizer, the availability of the nutrients was more.

In Kerala, FYM has been applied at the rate of 5 t/ha in all the seasons and the effect was studied. Coir dust was applied @ 500 kg/ ha. It showed that by applying FYM regularly the original dose of NPK for rice 70-35-35 kg/ha was reduced to 1/3rd of nitrogen and potassium and 2/3rd of phosphorus. Henceforth, the dose that was applied after the application of FYM in all the seasons was 47 kg N, 12 kg phosphorus and 23 kg. of potassium. The grain yield was also more in FYM treated plots.

Biogas Slurry in INM

Effect of biogas slurry was also studied in several vegetables for yield.

Table: *Effect of Biogas and chemical fertilizer combination on Vegetable yield*

Sl.No	*Crop*	*Yield from Biogas Slurry + CF (kg/plot)*	*Yield from Chemical fertilizer (kg/plot)*
1	Tomato	296	230
2	Cauliflower	528	492
3	Cabbage	177	152
4	Knol Khol	859	785
5	Capsicum	28	25
6	Onion	174	160

Conclusion

There are unending examples for integrated nutrient management which is difficult to mention all of them. Experiments are being conducted depending on the area, organic source availability in that area, crops and climatic conditions. They vary from place to place. It would be good if farmers realizes the need of organic matter to their soils and adopts this integrated nutrient management for sustaining the fertility of the soil.

Photosynthesis

Photosynthesis is a process used by plants and other organisms to convert light energy, normally from the sun, into chemical energy that can be used to fuel the organisms' activities. Carbohydrates, such as sugars, are synthesized from carbon dioxide and water. Oxygen is also released, mostly as a waste product. Most plants, most algae, and cyanobacteria perform the process of photosynthesis, and are called photoautotrophs. Photosynthesis maintains atmospheric oxygen levels and supplies all of the organic compounds and most of the energy necessary for all life on Earth.

Although photosynthesis is performed differently by different species, the process always begins when energy from light is absorbed by proteins called reaction centres that contain green chlorophyll pigments. In plants, these proteins are held inside organelles called chloroplasts, which are most abundant in leaf cells, while in bacteria they are embedded in the plasma membrane. In these light-dependent reactions, some energy is used to strip electrons from suitable substances such as water, producing oxygen gas. Furthermore, two further compounds are generated: reduced nicotinamide adenine dinucleotide phosphate (NADPH) and adenosine triphosphate (ATP), the "energy currency" of cells.

In plants, algae and cyanobacteria, sugars are produced by a subsequent sequence of light-independent reactions called the Calvin cycle, but some bacteria use different mechanisms, such as the reverse Krebs cycle. In the Calvin cycle, atmospheric carbon dioxide is incorporated into already existing organic carbon compounds, such as ribulose bisphosphate (RuBP). Using the ATP and NADPH produced by the light-dependent reactions, the resulting compounds are then reduced and removed to form further carbohydrates such as glucose.

The first photosynthetic organisms probably evolved early in the evolutionary history of life and most likely used reducing agents such as hydrogen or hydrogen sulfide as sources of electrons, rather than water. Cyanobacteria appeared later, and the excess oxygen they produced contributed to the oxygen catastrophe, which rendered the evolution of complex life possible. Today, the average rate of energy capture by photosynthesis globally is approximately 130 terawatts, which is about six times larger than the current power consumption of human civilization. Photosynthetic organisms also convert around 100–115 thousand million metric tonnes of carbon into biomass per year.

Overview

Photosynthetic organisms are photoautotrophs, which means that they are able to synthesize food directly from carbon dioxide and water using energy from light. However, not all organisms that use light as a source of energy carry out photosynthesis, since *photoheterotrophs* use organic compounds, rather than carbon dioxide, as a source of carbon. In plants, algae and cyanobacteria, photosynthesis releases oxygen. This is called *oxygenic photosynthesis*. Although there are some differences between oxygenic photosynthesis in plants, algae, and cyanobacteria, the overall process is quite similar in these organisms. However, there are some types of bacteria that carry out anoxygenic photosynthesis, which consumes carbon dioxide but does not release oxygen.

Carbon dioxide is converted into sugars in a process called carbon fixation. Carbon fixation is an endothermic redox reaction, so photosynthesis needs to supply both a source of energy to drive this process, and the electrons needed to convert carbon dioxide into a carbohydrate. This addition of the electrons is a reduction reaction. In general outline and in effect, photosynthesis is the opposite of cellular respiration, in which glucose and other compounds are oxidized to produce carbon dioxide and water, and to release exothermic chemical energy to drive the organism's metabolism. However, the two processes take place through a different sequence of chemical reactions and in different cellular compartments.

The general equation for photosynthesis is therefore:

$$2n\ CO_2 + 2n\ DH_2 + \text{photons} \rightarrow 2(CH_2O)_n + 2n\ DO$$

Carbon dioxide + electron donor + light energy → carbohydrate + oxidized electron donor

In *oxygenic* photosynthesis water is the electron donor and, since its hydrolysis releases oxygen, the equation for this process is:

$$2n\ CO_2 + 4n\ H_2O + \text{photons} \rightarrow 2(CH_2O)_n + 2n\ O_2 + 2n\ H_2O$$

carbon dioxide + water + light energy → carbohydrate + oxygen + water

Often 2n water molecules are cancelled on both sides, yielding:

$$2n\ CO_2 + 2n\ H_2O + \text{photons} \rightarrow 2(CH_2O)_n + 2n\ O_2$$

carbon dioxide + water + light energy → carbohydrate + oxygen

Other processes substitute other compounds (such as arsenite) for water in the electron-supply role; for example some microbes use sunlight to oxidize arsenite to arsenate: The equation for this reaction is:

$$CO_2 + (AsO_3^{3-}) + \text{photons} \rightarrow (AsO_4^{3-}) + CO$$

carbon dioxide + arsenite + light energy → arsenate + carbon monoxide (used to build other compounds in subsequent reactions)

Photosynthesis occurs in two stages. In the first stage, *light-dependent reactions* or *light reactions* capture the energy of light and use it to make the energy-storage molecules ATP and NADPH. During the second stage, the *light-independent reactions* use these products to capture and reduce carbon dioxide.

Most organisms that utilize photosynthesis to produce oxygen use visible light to do so, although at least three use shortwave infrared or, more specifically, far-red radiation.

Chloroplast

Chloroplasts are organelles, specialised subunits, in plant and algal cells. Their main role is to conduct photosynthesis, where the photosynthetic pigment chlorophyll captures the energy from sunlight, and stores it in the energy storage molecules ATP and NADPH while freeing oxygen from water. They then use the ATP and NADPH to make organic molecules from carbon dioxide in a process known as the Calvin cycle. Chloroplasts carry out a number of other functions, including fatty acid synthesis, much amino acid synthesis, and the immune response in plants.

A chloroplast is one of three types of plastid, characterized by its high concentration of chlorophyll. (The other two types, the leucoplast and the chromoplast), contain little chlorophyll and do not carry out photosynthesis.) Chloroplasts are highly dynamic—they circulate and are moved around within plant cells, and occasionally pinch in two to reproduce. Their behaviour is strongly influenced by environmental factors like light colour and intensity. Chloroplasts, like mitochondria, contain their own DNA, which is thought to be inherited from their ancestor—a photosynthetic cyanobacterium that was engulfed by an early eukaryotic cell. Chloroplasts cannot be made by the plant cell, and must be inherited by each daughter cell during cell division.

With one exception (a member of the genus *Paulinella*), all chloroplasts can probably be traced back to a single endosymbiotic event (the cyanobacterium being engulfed by the eukaryote). Despite this, chloroplasts can be found in an extremely wide set of organisms, some not even directly related to each other—a consequence of many secondary and even tertiary endosymbiotic events.

Chloroplast Lineages and Evolution

Chloroplasts are one of many types of organelles in the plant cell. They are considered to have originated from cyanobacteria through endosymbiosis—when a eukaryotic cell engulfed a photosynthesizing cyanobacterium which remained and became a permanent resident in the cell. Mitochondria are thought to have come from a similar event, where an ærobic prokaryote was engulfed. This origin of chloroplasts was first suggested by Konstantin Mereschkowski in 1905 after Andreas Schimper observed that chloroplasts closely resemble cyanobacteria in 1883. Chloroplasts are only found in plants and algae.

Cyanobacterial Ancestor

Cyanobacteria are considered the ancestors of chloroplasts. They are sometimes called blue-green algae even though they are prokaryotes. They are a diverse phylum of bacteria capable of carrying out photosynthesis, and are gram-negative, meaning they have two cell membranes. They also contain a peptidoglycan cell wall, which is thicker than in other gram-negative bacteria, and which is located between their two cell membranes. Like chloroplasts, they have thylakoids inside of them. On the thylakoid membranes are photosynthetic pigments, including chlorophyll *a*. Phycobilins are also common cyanobacterial pigments, usually organized into hemispherical phycobilisomes attached to the outside of the thylakoid membranes (phycobilins are not shared with all chloroplasts though).

Both chloroplasts and cyanobacteria have a double membrane, DNA, ribosomes, and thylakoids. Both the chloroplast and cyanobacterium depicted are idealized versions (the chloroplast is that of a higher plant)—a lot of diversity exists among chloroplasts and cyanobacteria.

Primary Endosymbiosis

A eukaryote with mitochondria engulfed a cyanobacterium in an event of serial primary endosymbiosis, creating a lineage of cells with both organelles. It is important to note that the cyanobacterial endosymbiont already had a double membrane—the phagosomal vacuole-derived membrane was lost.

Somewhere around a billion years ago, a free-living cyanobacterium entered an early eukaryotic cell, either as food or an internal parasite, and managed to escape the phagocytic vacuole it was contained in. The two innermost lipid-bilayer membranes that surround all chloroplasts correspond to the outer and inner membranes of the ancestral cyanobacterium's gram negative cell wall, and not the phagosomal membrane from the host, which was probably lost. The new cellular resident quickly became an advantage, providing food for the eukaryotic host, which allowed it to live within it. Over time, the cyanobacterium was assimilated, and many of its genes were lost or transferred to the nucleus of the host. Some of its proteins were then synthesized in the cytoplasm of the host cell, and imported back into the chloroplast.

This event is called *endosymbiosis*, or "cell living inside another cell". The cell living inside the other cell is called the *endosymbiont*; the endosymbiont is found inside the *host cell.*

Chloroplasts are believed to have arisen after mitochondria, since all eukaryotes contain mitochondria, but not all have chloroplasts. This is called *serial endosymbiosis*—an early eukaryote engulfing the mitochondrion ancestor, and some descendants of it then engulfing the chloroplast ancestor, creating a cell with both chloroplasts and mitochondria.

Whether or not chloroplasts came from a single endosymbiotic event, or many independent engulfments across various eukaryotic lineages has been long debated, but it is now generally held that all organisms with chloroplasts either share a single ancestor or obtained their chloroplast from organisms that share a common ancestor that took in a cyanobacterium 600–1600 million years ago.

These chloroplasts, which can be traced back directly to a cyanobacterial ancestor are known as *primary plastids* (*"plastid"* in this context means the almost the same thing as chloroplast). All primary chloroplasts belong to one of three chloroplast lineages—the glaucophyte chloroplast lineage, the rhodophyte, or red algal chloroplast lineage, or the chloroplastidan, or green chloroplast lineage. The second two are the largest, and the green chloroplast lineage is the one that contains the land plants.

Glaucophyta

The alga *Cyanophora*, a glaucophyte, is thought to be one of the first organisms to contain a chloroplast. The glaucophyte chloroplast group is the smallest of the three primary chloroplast lineages, being found in only thirteen species, and is thought to be the one that branched off the earliest. Glaucophytes have chloroplasts which retain a peptidoglycan wall between their double membranes, like their cyanobacterial parent. For this reason, glaucophyte chloroplasts are also known as *muroplasts*. Glaucophyte chloroplasts also contain concentric unstacked thylakoids which surround a carboxysome, an icosahedral structure that glaucophyte chloroplasts and cyanobacteria keep their carbon fixation enzyme rubisco in. The starch they synthesize collects outside the chloroplast. Like cyanobacteria, glaucophyte chloroplast thylakoids are studded with light collecting structures called phycobilisomes. For these reasons, glaucophyte chloroplasts are considered a primitive intermediate between cyanobacteria and the more evolved chloroplasts in red algae and plants.

Rhodophyceæ (Red Algae)

The rhodophyte, or red algal chloroplast group is another large and diverse chloroplast lineage. Rhodophyte chloroplasts are also called *rhodoplasts*, literally "red chloroplasts".

Rhodoplasts have a double membrane with an intermembrane space and phycobilin pigments organized into phycobilisomes on the thylakoid membranes, preventing their thylakoids from stacking. Some contain pyrenoids. Rhodoplasts have chlorophyll *a* and phycobilins for photosynthetic pigments; the phycobilin phycoerytherin is responsible for giving many red algae their distinctive red colour. However, since they also contain the blue-green chlorophyll *a* and other pigments, many are reddish to purple from the combination. The red phycoerytherin pigment is an adaptation to help red algae catch more sunlight in deep water—as such, some red algae that live in shallow water have less phycoerytherin in their rhodoplasts, and can appear more greenish. Rhodoplasts synthesize a form of starch called floridean, which collects into granules outside the rhodoplast, in the cytoplasm of the red alga.

Chloroplastida (Green Algae and Plants)

The chloroplastidan chloroplasts, or green chloroplasts, are another large, highly diverse primary chloroplast lineage. Their host organisms are commonly known as the green algae and land plants. They differ from glaucophyte and red algal chloroplasts in that they have lost their phycobilisomes, and contain chlorophyll *b* instead. Most green chloroplasts are (obviously) green, though some aren't, like some forms of *Hœmatococcus pluvialis*, due to accessory pigments that override the chlorophylls' green colours. Chloroplastidan chloroplasts have lost the peptidoglycan wall between their double membrane, and have replaced it with an intermembrane space. Some plants seem to have kept the genes for the synthesis of the peptidoglycan layer, though they've been repurposed for use in chloroplast division instead.

Most of the chloroplasts depicted in this article are green chloroplasts. Green algae and plants keep their starch *inside* their chloroplasts, and in plants and some algae, the chloroplast thylakoids are arranged in grana stacks. Some green algal chloroplasts contain a structure called a pyrenoid, which is functionally similar to the glaucophyte carboxysome in that it is where rubisco and CO_2 are concentrated in the chloroplast.

Helicosproidia

The helicosproidia are nonphotosynthetic parasitic green algae that are thought to contain a vestigal chloroplast. Genes from a chloroplast and nuclear genes indicating the presence of a chloroplast have been found in helicosporoidia. even if nobody's seen the chloroplast itself.

Secondary and Tertiary Endosymbiosis

Many other organisms obtained chloroplasts from the primary chloroplast lineages through secondary endosymbiosis—engulfing a red or green alga that contained a chloroplast. These chloroplasts are known as secondary plastids.

While primary chloroplasts have a double membrane from their cyanobacterial ancestor, secondary chloroplasts have additional membranes outside of the original two, as a result of the secondary endosymbiotic event, when a nonphotosynthetic eukaryote engulfed an chloroplast-containing alga but failed to digest it—much like the cyanobacterium at the beginning of this story. The engulfed alga was broken down, leaving only its chloroplast, and sometimes its cell membrane and nucleus, forming a chloroplast with three to four membranes—the two cyanobacterial membranes, sometimes the eaten alga's cell membrane, and the phagosomal vacuole from the host's cell membrane.

The genes in the phagocytosed eukaryote's nucleus are often transferred to the secondary host's nucleus. Cryptomonads and chlorarachniophytes retain the phagocytosed eukaryote's nucleus, an object called a nucleomorph, located between the second and third membranes of the chloroplast.

All secondary chloroplasts come from green and red algae—no secondary chloroplasts from glaucophytes have been observed, probably because glaucophytes are relatively rare in nature, making them less likely to have been taken up by another eukaryote.

Green Algal Derived Chloroplasts

Green algae have been taken up by the euglenids, chlorarachniophytes, a lineage of dinoflagellates, and possibly the ancestor of the chromalveolates in three or four separate engulfments. Many green algal derived chloroplasts contain pyrenoids, but unlike chloroplasts in their green algal ancestors, starch collects in granules outside the chloroplast.

Euglenophytes

Euglenophytes are a group of common flagellated protists that contain chloroplasts derived from a green alga. Euglenophyte chloroplasts have three membranes—it is thought that the membrane of the primary endosymbiont was lost, leaving the cyanobacterial membranes, and the secondary host's phagosomal membrane. Euglenophyte chloroplasts have a pyrenoid and thylakoids stacked in groups of three. Starch is stored in the form of paramylon, which is contained in membrane-bound granules in the cytoplasm of the euglenophyte.

Chlorarachniophytes

Chlorarachniophytes are a rare group of organisms that also contain chloroplasts derived from green algae, though their story is more complicated than that of the euglenophytes. The ancestor of chlorarachniophytes is thought to have been a chromalveolate, a eukaryote with a *red* algal derived chloroplast. It is then thought to have lost its first red algal chloroplast, and later engulfed a green alga, giving it its second, green algal derived chloroplast.

Chlorarachniophyte chloroplasts are bounded by four membranes, except near the cell membrane, where the chloroplast membranes fuse into a double membrane. Their thylakoids are arranged in loose stacks of three. Chlorarachniophytes have a form of starch called chrysolaminarin, which they store in the cytoplasm, often collected around the chloroplast pyrenoid, which bulges into the cytoplasm.

Chlorarachniophyte chloroplasts are notable because the green alga they are derived from has not been completely broken down—its nucleus still persists as a nucleomorph found between the second and third chloroplast membranes—the periplastid space, which corresponds to the green alga's cytoplasm.

Early Chromalveolates

Recent research has suggested that the ancestor of the chromalveolates acquired a green algal prasinophyte endosymbiont. The green algal derived chloroplast was lost and replaced with a red algal derived chloroplast, but not before contributing some of its genes to the early chromalveolate's nucleus. The presence of both green algal and red algal genes in chromalveolates probably helps them thrive under fluctuating light conditions.

Red Algal Derived Chloroplasts (Chromalveolate Chloroplasts)

Like green algae, red algae have also been taken up in secondary endosymbiosis, though it is thought that all red algal derived chloroplasts are descended from a single red alga that was engulfed by an early chromalveolate, giving rise to the chromalveolates, some of which, like the ciliates, subsequently lost the chloroplast. This is still debated though.

Pyrenoids and stacked thylakoids are common in chromalveolate chloroplasts, and the outermost membrane of many are continuous with the rough endoplasmic reticulum and studded with ribosomes. They have lost their phycobilisomes and exchanged them for chlorophyll *c*, which isn't found in primary red algal chloroplasts themselves.

Cryptophytes

Cryptophytes, or cryptomonads are a group of algae that contain a red-algal derived chloroplast. Cryptophyte chloroplasts contain a nucleomorph that superficially resembles that of the chlorarachniophytes. Cryptophyte chloroplasts have four membranes, the outermost of which is continuous with the rough endoplasmic reticulum. They synthesize ordinary starch, which is stored in granules found in the periplastid space—outside the original double membrane, in the place that corresponds to the red alga's cytoplasm. Inside cryptophyte chloroplasts is a pyrenoid and thylakoids in stacks of two.

Their chloroplasts do not have phycobilisomes, but they do have phycobilin pigments which they keep in their thylakoid space, rather than anchored on the outside of their thylakoid membranes.

Haptophytes

Haptophytes are similar and closely related to cryptophytes, and are thought to be the first chromalveolates to branch off. Their chloroplasts lack a nucleomorph, their thylakoids are in stacks of three, and they synthesize chrysolaminarin sugar, which they store completely outside of the chloroplast, in the cytoplasm of the haptophyte.

Heterokontophytes (Stramenopiles)

The heterokontophytes, also known as the stramenopiles, are a very large and diverse group of algae that also contain red algal derived chloroplasts. Heterokonts include the diatoms and the brown algae, golden algae, and yellow-green algae.

Heterokont chloroplasts are very similar to haptophyte chloroplasts, containing a pyrenoid, triplet thylakoids, and with some exceptions, having an epiplastid membrane connected to the endoplasmic reticulum. Like haptophytes, heterokontophytes store sugar in chrysolaminarin granules in the cytoplasm. Heterokontophyte chloroplasts contain chlorophyll *a* and with a few exceptions chlorophyll *c*, but also have carotenoids which give them their many colours.

Possible cladogram of chloroplast evolution

Apicomplexans: Apicomplexans are another group of chromalveolates. Like the helicosproidia, they're parasitic, and have a nonphotosynthetic chloroplast. They were once thought to be related to the helicosproidia, but it is now known that the helicosproida are green algae rather than chromalveolates. The apicomplexans include

Plasmodium, the malaria parasite. Many apicomplexans keep a vestigial red algal derived chloroplast called an apicoplast, which they inherited from their ancestors. Other apicomplexans like *Cryptosporidium* have lost the chloroplast completely. Apicomplexans store their energy in amylopectin starch granules that are located in their cytoplasm, even though they are nonphotosynthetic.

Apicoplasts have lost all photosynthetic function, and contain no photosynthetic pigments or true thylakoids. They are bounded by four membranes, but the membranes are not connected to the endoplasmic reticulum. The fact that apicomplexans still keep their nonphotosynthetic chloroplast around demonstrates how the chloroplast carries out important functions other than photosynthesis. Plant chloroplasts provide plant cells with many important things besides sugar, and apicoplasts are no different—they synthesize fatty acids, isopentenyl pyrophosphate, iron-sulphur clusters, and carry out part of the heme pathway. This makes the apicoplast an attractive target for drugs to cure apicomplexan-related diseases. The most important apicoplast function is isopentenyl pyrophosphate synthesis—in fact, apicomplexans die when something interferes with this apicoplast function, and when apicomplexans are grown in an isopentenyl pyrophosphate-rich medium, they dump the organelle.

Dinophytes

The dinoflagellates are yet another very large and diverse group of protists, around half of which are (at least partially) photosynthetic.

Most dinophyte chloroplasts are secondary red algal derived chloroplasts, like other chromalveolate chloroplasts. Many other dinophytes have lost the chloroplast (becoming the nonphotosynthetic kind of dinoflagellate), or replaced it though *tertiary* endosymbiosis—the engulfment of another chromalveolate containing a red algal derived chloroplast. Others replaced their original chloroplast with a green algal derived one. Most dinophyte chloroplasts contain at least the photosynthetic pigments chlorophyll *a*, chlorophyll c_2, *beta*-carotene, and at least one dinophyte-unique xanthophyll (peridinin, dinoxanthin, or diadinoxanthin), giving many a golden-brown colour. All dinophytes store starch in their cytoplasm, and most have chloroplasts with thylakoids arranged in stacks of three.

Peridinin-containing Dinophyte Chloroplast

The most common dinophyte chloroplast is the peridinin-type chloroplast, characterized by the carotenoid pigment peridinin in their chloroplasts, along with chlorophyll *a* and chlorophyll c_2. Peridinin is not found in any other group of chloroplasts. The peridinin chloroplast is bounded by three membranes (occasionally two), having lost the red algal endosymbiont's original cell membrane. The outermost membrane is not connected to the endoplasmic reticulum. They contain a pyrenoid, and have triplet-stacked thylakoids. Starch is found outside the chloroplast An important feature of these chloroplasts is that their chloroplast DNA is highly reduced and fragmented into many small circles. Most of the genome has migrated to the nucleus, and only critical photosynthesis-related genes remain in the chloroplast.

The peridinin chloroplast is thought to be the dinophytes' "original" chloroplast, which has been lost, reduced, replaced, or has company in several other dinophyte lineages.

Fucoxanthin-containing Dinophyte Chloroplasts (Haptophyte Endosymbionts)

The fucoxanthin dinophyte lineages (including *Karlodinium* and *Karenia*) lost their original red algal derived chloroplast, and replaced it with a new chloroplast derived from a haptophyte endosymbiont. *Karlodinium* and *Karenia* probably took up different heterokontophytes. Because the haptophyte chloroplast has four membranes, tertiary endosymbiosis would be expected to create a six membraned chloroplast, adding the haptophyte's cell membrane and the dinophyte's phagosomal vacuole. However, the haptophyte was heavily reduced, stripped of a few membranes and its nucleus, leaving only its chloroplast (with its original double membrane), and possibly one or two additional membranes around it.

Fucoxanthin-containing chloroplasts are characterized by having the pigment fucoxanthin (actually 192 -hexanoyloxy-fucoxanthin and/ or 192 -butanoyloxy-fucoxanthin) and no peridinin. Fucoxanthin is also found in haptophyte chloroplasts, providing evidence of ancestry.

Cryptophyte Derived Dinophyte Chloroplast

Members of the genus *Dinophysis* have a phycobilin-containing chloroplast taken from a cryptophyte. However, the cryptophyte is not an endosymbiont—only the chloroplast seems to have been taken, and the chloroplast has been stripped of its nucleomorph and outermost two membranes, leaving just a two-membraned chloroplast. Cryptophyte chloroplasts require their nucleomorph to maintain themselves, and *Dinophysis* species grown in cell culture alone cannot survive, so it is possible (but not confirmed) that the *Dinophysis* chloroplast is a kleptoplast—if so, *Dinophysis* chloroplasts wear out and *Dinophysis* species must continually engulf cryptophytes to obtain new chloroplasts to replace the old ones.

Diatom Derived Dinophyte Chloroplasts

Some dinophytes, like *Kryptoperidinium* and *Durinskia* have a diatom (heterokontophyte) derived chloroplast. These chloroplasts are bounded by up to *five* membranes, (depending on whether you count the entire diatom endosymbiont as the chloroplast, or just the red algal derived chloroplast inside it). The diatom endosymbiont has been reduced relatively little—it still retains its original mitochondria, and has endoplasmic reticulum, ribosomes, a nucleus, and of course, red algal derived chloroplasts—practically a complete cell, all inside the host's endoplasmic reticulum lumen. However the diatom endosymbiont can't store its own food—its starch is found in granules in the dinophyte host's cytoplasm instead. The diatom endosymbiont's nucleus is present, but it probably can't be called a nucleomorph because it shows no sign of genome reduction, and might have even been *expanded.* Diatoms have been engulfed by dinoflagellates at least three times.

The diatom endosymbiont is bounded by a single membrane, inside it are chloroplasts with four membranes. Like the diatom endosymbiont's diatom ancestor, the chloroplasts have triplet thylakoids and pyrenoids.

In some of these genera, the diatom endosymbiont's chloroplasts aren't the only chloroplasts in the dinophyte. The original three-membraned peridinin chloroplast is still around, converted to an eyespot.

Prasinophyte (Green Algal) Derived Dinophyte Chloroplast

Lepidodinium viride and its close relatives are dinophytes that lost their original peridinin chloroplast and replaced it with a green algal derived chloroplast (more specifically, a prasinophyte). *Lepidodinium* is the only dinophyte that has a chloroplast that's not from the rhodoplast lineage. The chloroplast is surrounded by two membranes and has no nucleomorph—all the nucleomorph genes have been transferred to the dinophyte nucleus. The endosymbiotic event that led to this chloroplast was serial secondary endosymbiosis rather than tertiary endosymbiosis—the endosymbiont was a green alga containing a primary chloroplast (making a secondary chloroplast).

Chromatophores

While most chloroplasts originate from that first set of endosymbiotic events, *Paulinella chromatophora* is an exception, which has acquired a photosynthetic cyanobacterial endosymbiont more recently. It is not clear whether that symbiont is closely related to the ancestral chloroplast of other eukaryotes. Being in the early stages of endosymbiosis, *Paulinella chromatophora* can offer some insights into how chloroplasts evolved. *Paulinella* cells contain one or two sausage shaped blue-green photosynthesizing structures called chromatophores, descended from the cyanobacterium *Synechococcus*. Chromatophores cannot survive outside their host. Chromatophore DNA is about a million base pairs long, containing around 850 protein encoding genes—far less than the three million base pair *Synechococcus* genome, but much larger than the approximately 150,000 base pair genome of the more assimilated chloroplast. Chromatophores have transferred much less of their DNA to the nucleus of their host. About 0.3–0.8% of the nuclear DNA in *Paulinella* is from the chromatophore, compared with 11–14% from the chloroplast in plants.

Kleptoplastidy

In some groups of mixotrophic protists, like some dinoflagellates, chloroplasts are separated from a captured alga or diatom and used temporarily. These klepto chloroplasts may only have a lifetime of a few days and are then replaced.

Chloroplast DNA

Chloroplasts have their own DNA, often abbreviated as ctDNA, or cpDNA. It is also known as the plastome. Its existence was first proved in 1962, and first sequenced in 1986—when two Japanese research teams sequenced the chloroplast DNA of liverwort and tobacco. Since then, hundreds of chloroplast DNAs from various species have been sequenced, but they're mostly those of land plants and green algae—glaucophytes, red algae, and other algal groups are extremely underrepresented, potentially introducing some bias in views of "typical" chloroplast DNA structure and content.

Molecular Structure

Chloroplast DNA Interactive gene map of chloroplast DNA from *Nicotiana tabacum*. Segments with labels on the inside reside on the B strand of DNA, segments with labels on the outside are on the A strand. Notches indicate introns.

With few exceptions, most chloroplasts have their entire chloroplast genome combined into a single large ring, typically 120,000–170,000 base pairs long. They can have a contour length of around 30–60 micrometers, and have a mass of about 80–130 million daltons.

While usually thought of as a circular molecule, there is some evidence that chloroplast DNA molecules more often take on a linear shape.

Inverted Repeats

Many chloroplast DNAs contain two *inverted repeats*, which separate a long single copy section (LSC) from a short single copy section (SSC). While a given pair of inverted repeats are rarely completely identical, they are always very similar to each other, apparently resulting from concerted evolution.

The inverted repeats vary wildly in length, ranging from 4,000 to 25,000 base pairs long each and containing as few as four or as many as over 150 genes. Inverted repeats in plants tend to be at the upper end of this range, each being 20,000–25,000 base pairs long.

The inverted repeat regions are highly conserved among land plants, and accumulate few mutations. Similar inverted repeats exist in the genomes of cyanobacteria and the other two chloroplast lineages (glaucophyta and rhodophyceæ), suggesting that they predate the chloroplast, though some chloroplast DNAs have since lost or flipped the inverted repeats (making them direct repeats). It is possible that the inverted repeats help stabilize the rest of the chloroplast genome, as chloroplast DNAs which have lost some of the inverted repeat segments tend to get rearranged more.

Nucleoids

New chloroplasts may contain up to 100 copies of their DNA, though the number of chloroplast DNA copies decreases to about 15–20 as the chloroplasts age. They are usually packed into nucleoids which can contain several identical chloroplast DNA rings. Many nucleoids can be found in each chloroplast. In primitive red algae, the chloroplast DNA nucleoids are clustered in the centre of the chloroplast, while in green plants and green algae, the nucleoids are dispersed throughout the stroma.

Though chloroplast DNA is not associated with true histones, in red algae, similar proteins that tightly pack each chloroplast DNA ring into a nucleoid have been found.

DNA Replication

Gene Content and Protein Synthesis: The chloroplast genome most commonly includes around 100 genes which code for a variety of things, mostly to do with the protein pipeline and photosynthesis. As in prokaryotes, genes in chloroplast DNA are organized into operons. Interestingly though, unlike prokaryotic DNA molecules, chloroplast DNA molecules contain introns (plant mitochondrial DNAs do too, but not human mtDNAs).

Among land plants, the contents of the chloroplast genome are fairly similar.

Chloroplast Genome Reduction and Gene Transfer

Over time, many parts of the chloroplast genome were transferred to the nuclear genome of the host, a process called *endosymbiotic gene transfer*. As a result, the chloroplast genome is heavily reduced compared to that of free-living cyanobacteria. Chloroplasts may contain 60–100 genes whereas cyanobacteria often have more than 1500 genes in their genome.

Endosymbiotic gene transfer is how we know about the lost chloroplasts in many chromalveolate lineages. Even if a chloroplast is eventually lost, the genes it donated to the former host's nucleus persist, providing evidence for the lost chloroplast's existence. For example, while diatoms (a heterokontophyte) now have a red algal derived chloroplast, the presence of many green algal genes in the diatom nucleus provide evidence that the diatom ancestor (probably the ancestor of all chromalveolates too) had a green algal derived chloroplast at some point, which was subsequently replaced by the red chloroplast.

In land plants, some 11–14% of the DNA in their nuclei can be traced back to the chloroplast, up to 18% in *Arabidopsis*, corresponding to about 4,500 protein-coding genes. There have been a few recent transfers of genes from the chloroplast DNA to the nuclear genome in land plants.

Of the approximately three-thousand proteins found in chloroplasts, some 95% of them are encoded by nuclear genes. Many of the chloroplast's protein complexes consist of subunits from both the chloroplast genome and the host's nuclear genome. As a result, protein synthesis must be coordinated between the chloroplast and the nucleus. The chloroplast is mostly under nuclear control, though chloroplasts can also give out signals regulating gene expression in the nucleus, called *retrograde signalling*.

Protein Synthesis

Protein synthesis within chloroplasts relies on two RNA polymerases. One is coded by the chloroplast DNA, the other is of nuclear origin. The two RNA polymerases may recognize and bind to different kinds of promoters within the chloroplast genome. The ribosomes in chloroplasts are similar to bacterial ribosomes.

Protein Targeting and Import

The movement of so many chloroplast genes to the nucleus means that lots of chloroplast proteins that were supposed to be translated in the chloroplast are now synthesized in the cytoplasm. This means that these proteins must be directed back to the chloroplast, and imported through at least two chloroplast membranes.

Curiously, around half of the protein products of transferred genes aren't even targeted back to the chloroplast. Many became exaptations, taking on new functions like participating in cell division, protein routing, and even disease resistance. A few chloroplast genes found new homes in the mitochondrial genome—most became nonfunctional pseudogenes, though a few tRNA genes still work in the mitochondrion. Some transferred chloroplast DNA protein products get directed to the secretory pathway (though it should be noted that many secondary plastids are bounded by an outermost membrane derived from the host's cell membrane, and therefore topologically outside of the cell, because to reach the chloroplast from the cytosol, you have to cross the cell membrane, just like if you were headed for the extracellular space. In those cases, chloroplast-targeted proteins do initially travel along the secretory pathway).

Because the cell acquiring a chloroplast already had mitochondria (and peroxisomes, and a cell membrane for secretion), the new chloroplast host had to develop a unique protein targeting system to avoid having chloroplast proteins being sent to the wrong organelle, and the wrong proteins being sent to the chloroplast.

The two ends of a polypeptide are called the N-terminus, or *amino end*, and the C-terminus, or *carboxyl end*. This polypeptide has four amino acids linked together. At the left is the N-terminus, with its amino (H_2N) group in green. The blue C-terminus, with its carboxyl group (CO_2H) is at the right.

In most, but not all cases, nuclear-encoded chloroplast proteins are translated with a *cleavable transit peptide* that's added to the N-terminus of the protein precursor. Sometimes the transit sequence is found on the C-terminus of the protein, or within the functional part of the protein.

Transport Proteins and Membrane Translocons

After a chloroplast polypeptide is synthesized on a ribosome in the cytosol, an enzyme specific to chloroplast proteins phosphorylates, or adds a phosphate group to many (but not all) of them in their transit sequences. Phosphorylation helps many proteins bind the polypeptide, keeping it from folding prematurely. This is important because it prevents chloroplast proteins from assuming their active form and carrying out their chloroplast functions in the wrong place—the cytosol. At the same time, they have to keep just enough shape so that they can be recognized by the chloroplast. These proteins also help the polypeptide get imported into the chloroplast.

From here, chloroplast proteins bound for the stroma must pass through two protein complexes—the TOC complex, or *translocon on the outer chloroplast membrane*, and the TIC translocon, or *translocon on the inner chloroplast membrane translocon*. Chloroplast polypeptide chains probably often travel through the two complexes at the same time, but the TIC complex can also retrieve preproteins lost in the intermembrane space.

Structure

In land plants, chloroplasts are generally lens-shaped, 5–8 μm in diameter and 1–3 μm thick. Greater diversity in chloroplast shapes exists among the algae, which often contain a single chloroplast that can be shaped like a net (e.g., *Oedogonium*), a cup (e.g., *Chlamydomonas*), a ribbon-like spiral around the edges of the cell (e.g., *Spirogyra*), or slightly twisted bands at the cell edges (e.g., *Sirogonium*). Some algae have two chloroplasts in each cell; they are star-shaped in *Zygnema*, or may follow the shape of half the cell in order Desmidiales. In some algae, the chloroplast takes up most of the cell, with pockets for the nucleus and other organelles (for example some species of *Chlorella* have a cup-shaped chloroplast that occupies much of the cell).

All chloroplasts have at least three membrane systems—the outer chloroplast membrane, the inner chloroplast membrane, and the thylakoid system. Chloroplasts that are the product of secondary endosymbiosis may have additional membranes surrounding these three. Inside the outer and inner chloroplast membranes is the chloroplast stroma, a semi-gel-like fluid that makes up much of a chloroplast's volume, and in which the thylakoid system floats.

There are some common misconceptions about the outer and inner chloroplast membranes. The fact that chloroplasts are surrounded by a double membrane is often cited as evidence that they are the descendants of endosymbiotic cyanobacteria. This is often interpreted as meaning the outer chloroplast membrane is the product of the host's cell membrane infolding to form a vesicle to surround the ancestral cyanobacterium—which is not true—both chloroplast membranes are homologous to the cyanobacterium's original double membranes.

The chloroplast double membrane is also often compared to the mitochondrial double membrane. This is not a valid comparison—the inner mitochondria membrane is used to run proton pumps and carry out oxidative phosphorylation across to generate ATP energy. The only chloroplast structure that can considered analogous to it is the internal thylakoid system. Even so, in terms of "in-out", the direction of chloroplast H^+ ion flow is in the opposite direction compared to oxidative phosphorylation in mitochondria. In addition, in terms of function, the inner chloroplast membrane, which regulates metabolite passage and synthesizes some materials, has no counterpart in the mitochondrion.

Outer Chloroplast Membrane

The outer chloroplast membrane is a semi-porous membrane that small molecules and ions can easily diffuse across. However, it is not permeable to larger proteins, so chloroplast polypeptides being synthesized in the cell cytoplasm must be transported across the outer chloroplast membrane by the TOC complex, or *translocon on the outer chloroplast* membrane.

The chloroplast membranes sometimes protrude out into the cytoplasm, forming a stromule, or stroma-containing tubule. Stromules are very rare in chloroplasts, and are much more common in other plastids like chromoplasts and amyloplasts in petals and roots, respectively. They may exist to increase the chloroplast's surface area for cross-membrane transport, because they are often branched and tangled with the endoplasmic reticulum. They were once thought to connect chloroplasts allowing them to exchange proteins, however recent research strongly refutes this idea. Observed stromules are probably just oddly shaped chloroplasts with constricted regions or dividing chloroplasts.

Intermembrane Space and Peptidoglycan Wall

Usually, a thin intermembrane space about 10–20 nanometers thick exists between the outer and inner chloroplast membranes.

Glaucophyte algal chloroplasts have a peptidoglycan layer between the chloroplast membranes. It corresponds to the peptidoglycan cell wall of their cyanobacterial ancestors, which is located between their two cell membranes. These chloroplasts are called muroplasts (from Latin *"mura"*, meaning "wall"). Other chloroplasts have lost the cyanobacterial wall, leaving an intermembrane space between the two chloroplast envelope membranes.

Inner Chloroplast Membrane

The inner chloroplast membrane borders the stroma and regulates passage of materials in and out of the chloroplast. After passing through the TOC complex in the outer chloroplast membrane, polypeptides must pass through the TIC complex *(translocon on the inner chloroplast membrane)* which is located in the inner chloroplast membrane.

In addition to regulating the passage of materials, the inner chloroplast membrane is where fatty acids, lipids, and carotenoids are synthesized.

Peripheral Reticulum

Some chloroplasts contain a structure called the chloroplast peripheral reticulum. It is often found in the chloroplasts of C_4 plants, though it has also been found in some C_3 angiosperms, and even some gymnosperms. The chloroplast peripheral reticulum consists of a maze of membranous tubes and vesicles continuous with the inner chloroplast membrane that extends into the internal stromal fluid of the chloroplast. Its purpose is thought to be to increase the chloroplast's surface area for cross-membrane transport between its stroma and the cell cytoplasm. The small vesicles sometimes observed may serve as transport vesicles to shuttle stuff between the thylakoids and intermembrane space.

Stroma

The protein-rich, alkaline, aqueous fluid within the inner chloroplast membrane and outside of the thylakoid space is called the stroma, which corresponds to the cytosol of the original cyanobacterium. Nucleoids of chloroplast DNA, chloroplast ribosomes, the thylakoid system with plastoglobuli, starch granules, and many proteins can be found floating around in it. The Calvin cycle, which fixes CO_2 into sugar takes place in the stroma.

Chloroplast Ribosomes

Chloroplasts have their own ribosomes, which they use to synthesize a small fraction of their proteins. Chloroplast ribosomes are about two-thirds the size of cytoplasmic ribosomes (around 17 nm vs 25 nm). They take mRNAs transcribed from the chloroplast DNA and translate them into protein. While similar to bacterial ribosomes, chloroplast translation is more complex than in bacteria, so chloroplast ribosomes include some chloroplast-unique features.

Plastoglobuli

Plastoglobuli (singular *plastoglobulus*, sometimes spelled *plastoglobule(s)*), are spherical bubbles of lipids and proteins about 45–60 nanometers across. They are surrounded by a lipid monolayer. Plastoglobuli are found in all chloroplasts, but become more common when the chloroplast is under oxidative stress, or when it ages and transitions into a gerontoplast. Plastoglobuli also exhibit a greater size variation under these conditions. They are also common in etioplasts, but decrease in number as the etioplasts mature into chloroplasts.

Plastoglubuli contain both structural proteins and enzymes involved in lipid synthesis and metabolism. They contain many types of lipids including plastoquinone, vitamin E, carotenoids and chlorophylls.

Plastoglobuli were once thought to be free-floating in the stroma, but it is now thought that they are permanently attached either to a thylakoid or to another plastoglobulus attached to a thylakoid, a configuration that allows a plastoglobulus to exchange its contents with the thylakoid network. In normal green chloroplasts, the vast majority of plastoglobuli occur singularly, attached directly to their parent thylakoid. In old or stressed chloroplasts, plastoglobuli tend to occur in linked groups or chains, still always anchored to a thylakoid.

Plastoglobuli form when a bubble appears between the layers of the lipid bilayer of the thylakoid membrane, or bud from existing plastoglubuli—though they never detach and float off into the stroma. Practically all plastoglobuli form on or near the highly curved edges of the thylakoid disks or sheets. They are also more common on stromal thylakoids than on granal ones.

Starch Granules

Starch granules are very common in chloroplasts, typically taking up 15% of the organelle's volume, though in some other plastids like amyloplasts, they can be big enough to distort the shape of the organelle. Starch granules are simply accumulations of starch in the stroma, and are not bounded by a membrane.

Starch granules appear and grow throughout the day, as the chloroplast synthesizes sugars, and are consumed at night to fuel respiration and continue sugar export into the phloem, though in mature chloroplasts, it is rare for a starch granule to be completely consumed or for a new granule to accumulate.

Starch granules vary in composition and location across different chloroplast lineages. In red algae, starch granules are found in the cytoplasm rather than in the chloroplast. In C_4 plants, mesophyll chloroplasts, which do not synthesize sugars, lack starch granules.

Rubisco

The chloroplast stroma contains many proteins, though the most common and important is Rubisco, which is probably also the most abundant protein on the planet. Rubisco is the enzyme that fixes CO_2 into sugar molecules. In C_3 plants, rubisco is abundant in all chloroplasts, though in C_4 plants, it is confined to the bundle sheath chloroplasts, where the Calvin cycle is carried out in C_4 plants.

Pyrenoids

The chloroplasts of some hornworts and algae contain structures called pyrenoids. They are not found in higher plants. Pyrenoids are roughly spherical and highly refractive bodies which are a site of starch accumulation in plants that contain them. They consist of an matrix opaque to electrons, surrounded by two hemispherical starch plates. The starch is accumulated as the pyrenoids mature. In algae with carbon concentrating mechanisms, the enzyme rubisco is found in the pyrenoids. Starch can also accumulate around the pyrenoids when CO_2 is scarce. Pyrenoids can divide to form new pyrenoids, or be produced "de novo".

Thylakoid System

uspended within the chloroplast stroma is the thylakoid system, a highly dynamic collection of membranous sacks called thylakoids where chlorophyll is found and the light reactions of photosynthesis happen. In most vascular plant chloroplasts, the thylakoids are arranged in stacks called grana, though in certain C_4 plant chloroplasts and some algal chloroplasts, the thylakoids are free floating.

Granal Structure

Using a light microscope, it is just barely possible to see tiny green granules—which were named grana. With electron microscopy, it became possible to see the thylakoid system in more detail, revealing it to consist of stacks of flat thylakoids which made up the grana, and long interconnecting stromal thylakoids which linked different grana. In the transmission electron microscope, thylakoid membranes appear as alternating light-and-dark bands, 8.5 nanometers thick.

For a long time, the three dimensional structure of the thylakoid system has been unknown or disputed. One model has the granum as a stack of thylakoids linked by helical stromal thylakoids; the other has the granum as a single folded thylakoid connected in a "hub and spoke" way to other grana by stromal thylakoids. While the thylakoid system is still commonly depicted according to the folded thylakoid model, it was determined in 2011 that the stacked and helical thylakoids model is correct.

In the helical thylakoid model, grana consist of a stack of flattened circular granal thylakoids that resemble pancakes. Each granum can contain anywhere from two to a hundred thylakoids, though grana with 10–20 thylakoids are most common. Wrapped around the grana are helicoid stromal thylakoids, also known as frets or lamellar thylakoids. The helices ascend at an angle of 20–25°, connecting to each granal thylakoid at a bridge-like slit junction. The helicoids may extend as large sheets that link multiple grana, or narrow to tube-like bridges between grana. While different parts of the thylakoid system contain different membrane proteins, the thylakoid membranes are continuous and the thylakoid space they enclose form a single continuous labyrinth.

Thylakoids

Thylakoids (sometimes spelled *thylakoïds*), are small interconnected sacks which contain the membranes that the light reactions of photosynthesis take place on. The word *thylakoid* comes from the Greek word *thylakos* which means "sack".

Embedded in the thylakoid membranes are important protein complexes which carry out the light reactions of photosynthesis. Photosystem II and photosystem I contain light-harvesting complexes with chlorophyll and carotenoids that absorb light energy and use it to energize electrons. Molecules in the thylakoid membrane use the energized electrons to pump hydrogen ions into the thylakoid space, decreasing the pH and turning it acidic. ATP synthase is a large protein complex that harnesses the concentration gradient of the hydrogen ions in the thylakoid space to generate ATP energy as the hydrogen ions flow back out into the stroma—much like a dam turbine.

There are two types of thylakoids—granal thylakoids, which are arranged in grana, and stromal thylakoids, which are in contact with the stroma. Granal thylakoids are pancake-shaped circular disks about 300–600 nanometers in diameter. Stromal thylakoids are helicoid sheets that spiral around grana. The flat tops and bottoms of granal thylakoids contain only the relatively flat photosystem II protein complex. This allows them to stack tightly, forming grana with many layers of tightly appressed membrane, called granal membrane, increasing stability and surface area for light capture.

In contrast, photosystem I and ATP synthase are large protein complexes which jut out into the stroma. They can't fit in the appressed granal membranes, and so are found in the stromal thylakoid membrane—the edges of the granal thylakoid disks and the stromal thylakoids. These large protein complexes may act as spacers between the sheets of stromal thylakoids.

The number of thylakoids and the total thylakoid area of a chloroplast is influenced by light exposure. Shaded chloroplasts contain larger and more grana with more thylakoid membrane area than chloroplasts exposed to bright light, which have smaller and fewer grana and less thylakoid area. Thylakoid extent can change within minutes of light exposure or removal.

Pigments and Chloroplast Colours

Inside the photosystems embedded in chloroplast thylakoid membranes are various photosynthetic pigments, which absorb and transfer light energy. The types of pigments found are different in various groups of chloroplasts, and are responsible for a wide variety of chloroplast colorations.

Chlorophylls

Chlorophyll *a* is found in all chloroplasts, as well as their cyanobacterial ancestors. Chlorophyll *a* is a blue-green pigment partially responsible for giving most cyanobacteria and chloroplasts their colour. Other forms of chlorophyll exist, such as the accessory pigments chlorophyll *b*, chlorophyll *c*, chlorophyll *d*, and chlorophyll *f*.

Chlorophyll *b* is an olive green pigment found only in the chloroplasts of plants, green algae, any secondary chloroplasts obtained through the secondary endosymbiosis of a green alga, and a few cyanobacteria. It is the chlorophylls *a* and *b* together that make most plant and green algal chloroplasts green.

Chlorophyll *c* is mainly found in secondary endosymbiotic chloroplasts that originated from a red alga, although it is not found in chloroplasts of red algae themselves. Chlorophyll *c* is also found in some green algae and cyanobacteria.

Chlorophylls *d* and *f* are pigments found only in some cyanobacteria.

Carotenoids

In addition to chlorophylls, another group of yellow–orange pigments called carotenoids are also found in the photosystems. There are about thirty photosynthetic carotenoids. They help transfer and dissipate excess energy, and their bright colours sometimes override the chlorophyll green, like during the fall, when the leaves of some land plants change colour. β-carotene is a bright red-orange carotenoid found in nearly all chloroplasts, like chlorophyll *a*. Xanthophylls, especially the orange-red zeaxanthin, are also common. Many other forms of carotenoids exist that are only found in certain groups of chloroplasts.

Phycobilins

Phycobilins are a third group of pigments found in cyanobacteria, and glaucophyte, red algal, and cryptophyte chloroplasts. Phycobilins come in all colours, though phycoerytherin is one of the pigments that makes many red algae red. Phycobilins often organize into relatively large protein complexes about 40 nanometers across called phycobilisomes. Like photosystem I and ATP synthase, phycobilisomes jut into the stroma, preventing thylakoid stacking in red algal chloroplasts. Cryptophyte chloroplasts and some cyanobacteria don't have their phycobilin pigments organized into phycobilisomes, and keep them in their thylakoid space instead.

Specialized Chloroplasts in C_4 Plants

To fix carbon dioxide into sugar molecules in the process of photosynthesis, chloroplasts use an enzyme called rubisco. Rubisco has a problem—it has trouble distinguishing between carbon dioxide and oxygen, so at high oxygen concentrations, rubisco starts accidentally adding oxygen to sugar precursors. This has the end result of ATP energy being wasted and CO_2 being released, all with no sugar being produced. This is a big problem, since O_2 is produced by the initial light reactions of photosynthesis, causing issues down the line in the Calvin cycle which uses rubisco.

C_4 plants evolved a way to solve this—by spatially separating the light reactions and the Calvin cycle. The light reactions, which store light energy in ATP and NADPH, are done in the mesophyll cells of a C_4 leaf. The Calvin cycle, which uses the stored energy to make sugar using rubisco, is done in the bundle sheath cells, a layer of cells surrounding a vein in a leaf.

As a result, chloroplasts in C_4 mesophyll cells and bundle sheath cells are specialized for each stage of photosynthesis. In mesophyll cells, chloroplasts are specialized for the light reactions, so they lack rubisco, and have normal grana and thylakoids, which they use to make ATP and NADPH, as well as oxygen. They store CO_2 in a four-carbon compound, which is why the process is called *C_4 photosynthesis.* The four-carbon compound is then transported to the bundle sheath chloroplasts, where it drops off CO_2 and returns to the mesophyll. Bundle sheath chloroplasts do not carry out the light reactions, preventing oxygen from building up in them and disrupting rubisco activity. Because of this, they lack thylakoids organized into grana stacks—though bundle sheath chloroplasts still have free-floating thylakoids in the stroma where they still carry out cyclic electron flow, a light-driven method of synthesizing ATP to power the Calvin cycle without generating oxygen. They lack photosystem II, and only have photosystem I—the only protein complex needed for cyclic electron flow. Because the job of bundle sheath chloroplasts is to carry out the Calvin cycle and make sugar, they often contain large starch grains.

Both types of chloroplast contain large amounts of chloroplast peripheral reticulum, which they use to get more surface area to transport stuff in and out of them. Mesophyll chloroplasts have a little more peripheral reticulum than bundle sheath chloroplasts.

Location

Distribution in a Plant

Not all cells in a multicellular plant contain chloroplasts. All green parts of a plant contain chloroplasts—the chloroplasts, or more specifically, the chlorophyll in them are what make the photosynthetic parts of a plant green. The plant cells which contain chloroplasts are usually parenchyma cells, though chloroplasts can also be found in collenchyma tissue. A plant cell which contains chloroplasts is known as a chlorenchyma cell. A typical chlorenchyma cell of a land plant contains about 10 to 100 chloroplasts.

In some plants such as cacti, chloroplasts are found in the stems, though in most plants, chloroplasts are concentrated in the leaves. One square millimeter of leaf tissue can contain half a million chloroplasts.

Within a leaf, chloroplasts are mainly found in the mesophyll layers of a leaf, and the guard cells of stomata. Palisade mesophyll cells can contain 30–70 chloroplasts per cell, while stomatal guard cells contain only around 8–15 per cell, as well as much less chlorophyll. Chloroplasts can also be found in the bundle sheath cells of a leaf, especially in C_4 plants, which carry out the Calvin cycle in their bundle sheath cells. They are often absent from the epidermis of a leaf.

Cellular Location

Chloroplast Movement: The chloroplasts of plant and algal cells can orient themselves to best suit the available light. In low-light conditions, they will spread out in a sheet—maximizing the surface area to absorb light. Under intense light, they will seek shelter by aligning in vertical columns along the plant cell's cell wall or turning sideways so that light strikes them edge-on. This reduces exposure and protects them from photooxidative damage. This ability to distribute chloroplasts so that they can take shelter behind each other or spread out may be the reason why land plants evolved to have many small chloroplasts instead of a few big ones. Chloroplast movement is considered one of the most closely regulated stimulus-response systems that can be found in plants. Mitochondria have also been observed to follow chloroplasts as they move.

In higher plants, chloroplast movement is run by phototropins, blue light photoreceptors also responsible for plant phototropism. In some algae, mosses, ferns, and flowering plants, chloroplast movement is influenced by red light in addition to blue light, though very long red wavelengths inhibit movement rather than speeding it up. Blue light generally causes chloroplasts to seek shelter, while red light draws them out to maximize light absorption. Studies of *Vallisneria gigantea*, an aquatic flowering plant, have shown that chloroplasts can get moving within five minutes of light exposure, though they don't initially show any net directionality. They may move along microfilament tracks, and the fact that the microfilament mesh changes shape to form a honeycomb structure surrounding the chloroplasts after they have moved suggests that microfilaments may help to anchor chloroplasts in place.

Function and Chemistry

Guard Cell Chloroplasts

Unlike most epidermal cells, the guard cells of plant stomata contain relatively well developed chloroplasts. However, exactly what they do is controversial.

Plant Innate Immunity

Plants lack specialized immune cells—all plant cells participate in the plant immune response. Chloroplasts, along with the nucleus, cell membrane, and endoplasmic reticulum, are key players in pathogen defence. Due to its role in a plant cell's immune response, pathogens frequently target the chloroplast.

Plants have two main immune responses—the hypersensitive response, in which infected cells seal themselves off and undergo programmed cell death, and systemic acquired resistance, where infected cells release signals warning the rest of the plant of a pathogen's presence. Chloroplasts stimulate both responses by purposely damaging their photosynthetic system, producing reactive oxygen species. High levels of reactive oxygen species will cause the hypersensitive response. The reactive oxygen species also directly kill any pathogens within the cell. Lower levels of reactive oxygen species initiate systemic acquired resistance, triggering defence-molecule production in the rest of the plant.

In some plants, chloroplasts are known to move closer to the infection site and the nucleus during an infection.

Chloroplasts can serve as cellular sensors. After detecting stress in a cell, which might be due to a pathogen, chloroplasts begin producing molecules like salicylic acid, jasmonic acid, nitric oxide and reactive oxygen species which can serve as defence-signals. As cellular signals, reactive oxygen species are unstable molecules, so they probably don't leave the chloroplast, but instead pass on their signal to an unknown second messenger molecule. All these molecules initiate retrograde signalling—signals from the chloroplast that regulate gene expression in the nucleus.

In addition to defence signalling, chloroplasts, with the help of the peroxisomes, help synthesize an important defence molecule, jasmonate. Chloroplasts synthesize all the fatty acids in a plant cell—linoleic acid, a fatty acid, is a precursor to jasmonate.

Photosynthesis

One of the main functions of the chloroplast is its role in photosynthesis, the process by which light is transformed into chemical energy, to subsequently produce food in the form of sugars. Water (H_2O) and carbon dioxide (CO_2) are used in photosynthesis, and sugar and oxygen (O_2) is made, using light energy. Photosynthesis is divided into two stages—the light reactions, where water is split to produce oxygen, and the dark reactions, or Calvin cycle, which builds sugar molecules from carbon dioxide. The two phases are linked by the energy carriers adenosine triphosphate (ATP) and nicotinamide adenine dinucleotide phosphate ($NADP^+$).

Light Reactions

The light reactions take place on the thylakoid membranes. They take light energy and store it in NADPH, a form of $NADP^+$, and ATP to fuel the dark reactions.

Energy Carriers

ATP is the phosphorylated version of adenosine diphosphate (ADP), which stores energy in a cell and powers most cellular activities. ATP is the energized form, while ADP is the (partially) depleted form. $NADP^+$ is an electron carrier which ferries high energy electrons. In the light reactions, it gets reduced, meaning it picks up electrons, becoming NADPH.

Photophosphorylation

Like mitochondria, chloroplasts use the potential energy stored in an H^+, or hydrogen ion gradient to generate ATP energy. The two photosystems capture light energy to energize electrons taken from water, and release them down an electron transport chain. The molecules between the photosystems harness the electrons' energy to pump hydrogen ions into the thylakoid space, creating a concentration gradient, with more hydrogen ions (up to a thousand times as many) inside the thylakoid system than in the stroma. The hydrogen ions in the thylakoid space then diffuse back down their concentration gradient, flowing back out into the stroma through ATP synthase. ATP synthase uses the energy from the flowing hydrogen ions to phosphorylate adenosine diphosphate into adenosine triphosphate, or ATP. Because chloroplast ATP synthase projects out into the stroma, the ATP is synthesized there, in position to be used in the dark reactions.

$NADP^+$ reduction

Electrons are often removed from the electron transport chains to charge $NADP^+$ with electrons, reducing it to NADPH. Like ATP synthase, ferredoxin-$NADP^+$ reductase, the enzyme that reduces $NADP^+$, releases the NADPH it makes into the stroma, right where it is needed for the dark reactions.

Because $NADP^+$ reduction removes electrons from the electron transport chains, they must be replaced—the job of photosystem II, which splits water molecules (H_2O) to obtain the electrons from its hydrogen atoms.

Cyclic Photophosphorylation

While photosystem II photolyzes water to obtain and energize new electrons, photosystem I simply reenergizes depleted electrons at the end of an electron transport chain. Normally, the reenergized electrons are taken by $NADP^+$, though sometimes they can flow back down more H^+-pumping electron transport chains to transport more hydrogen ions into the thylakoid space to generate more ATP. This is termed cyclic photophosphorylation because the electrons are recycled. Cyclic photophosphorylation is common in C_4 plants, which need more ATP than NADPH.

Dark Reactions

The Calvin cycle, also known as the dark reactions, is a series of biochemical reactions that fixes CO_2 into G3P sugar molecules and uses the energy and electrons from the ATP and NADPH made in the light reactions. The Calvin cycle takes place in the stroma of the chloroplast.

While named *"the dark reactions"*, in most plants, they take place in the light, since the dark reactions are dependent on the products of the light reactions.

Carbon Fixation and G3P Synthesis

The Calvin cycle starts by using the enzyme Rubisco to fix CO_2 into five-carbon Ribulose bisphosphate (RuBP) molecules. The result is unstable six-carbon molecules that immediately break down into three-carbon molecules called 3-phosphoglyceric acid, or 3-PGA. The ATP and NADPH made in the light reactions is used to convert the 3-PGA into glyceraldehyde-3-phosphate, or G3P sugar molecules. Most of the G3P molecules are recycled back into RuBP using energy from more ATP, but one out of every six produced leaves the cycle—the end product of the dark reactions.

Sugars and Starches

Glyceraldehyde-3-phosphate can double up to form larger sugar molecules like glucose and fructose. These molecules are processed, and from them, the still larger sucrose, a disaccharide commonly known as table sugar, is made, though this process takes place outside of the chloroplast, in the cytoplasm.

Sucrose is made up of a glucose monomer (left), and a fructose monomer (right).

Alternatively, glucose monomers in the chloroplast can be linked together to make starch, which accumulates into the starch grains found in the chloroplast. Under conditions such as high atmospheric CO_2 concentrations, these starch grains may grow very large, distorting the grana and thylakoids. The starch granules displace the thylakoids, but leave them intact. Waterlogged roots can also cause starch buildup in the chloroplasts, possibly due to less sucrose being exported out of the chloroplast (or more accurately, the plant cell). This depletes a plant's free phosphate supply, which indirectly stimulates chloroplast starch synthesis. While linked to low photosynthesis rates, the starch grains themselves may not necessarily interfere significantly with the efficiency of photosynthesis, and might simply be a side effect of another photosynthesis-depressing factor.

Photorespiration

Photorespiration can occur when the oxygen concentration is too high. Rubisco cannot distinguish between oxygen and carbon dioxide very well, so it can accidentally add O_2 instead of CO_2 to RuBP. This process reduces the efficiency of photosynthesis—it consumes ATP and oxygen, releases CO_2, and produces no sugar. It can waste up to half the carbon fixed by the Calvin cycle. Several mechanisms have evolved in different lineages that raise the carbon dioxide concentration relative to oxygen within the chloroplast, increasing the efficiency of photosynthesis. These mechanisms are called carbon dioxide concentrating mechanisms, or CCMs. These include Crassulacean acid metabolism, C_4 carbon fixation, and pyrenoids. Chloroplasts in C_4 plants are notable as they exhibit a distinct chloroplast dimorphism.

pH

Because of the H^+ gradient across the thylakoid membrane, the interior of the thylakoid is acidic, with a pH around 4, while the stroma is slightly basic, with a pH of around 8. The optimal stroma pH for the Calvin cycle is 8.1, with the reaction nearly stopping when the pH falls below 7.3.

CO_2 in water can form carbonic acid, which can disturb the pH of isolated chloroplasts, interfering with photosynthesis, even though CO_2 is used in photosynthesis. However, chloroplasts in living plant cells are not affected by this as much. Chloroplasts can pump K^+ and H^+ ions in and out of themselves using a poorly understood light-driven transport system.

In the presence of light, the pH of the thylakoid lumen can drop up to 1.5 pH units, while the pH of the stroma can rise by nearly one pH unit.

Amino Acid Synthesis

Chloroplasts alone make almost all of a plant cell's amino acids in their stroma except the sulphur-containing ones like cysteine and methionine. Cysteine is made in the chloroplast (the proplastid too) but it is also synthesized in the cytosol and mitochondria, probably because it has trouble crossing membranes to get to where it is needed. The chloroplast is known to make the precursors to methionine but it is unclear whether the organelle carries out the last leg of the pathway or if it happens in the cytosol.

Other Nitrogen Compounds

Chloroplasts make all of a cell's purines and pyrimidines—the nitrogenous bases found in DNA and RNA. They also convert nitrite (NO_2^-) into ammonia (NH_3) which supplies the plant with nitrogen to make its amino acids and nucleotides.

Other Chemical Products

Chloroplasts are the site of complex lipid metabolism.

7

Differentiation, Replication and Inheritance

Chloroplasts are a special type of a plant cell organelle called a plastid, though the two terms are sometimes used interchangeably. There are many other types of plastids, which carry out various functions. All chloroplasts in a plant are descended from undifferentiated proplastids found in the zygote, or fertilized egg. Proplastids are commonly found in an adult plant's apical meristems. Chloroplasts do not normally develop from proplastids in root tip meristems—instead, the formation of starch-storing amyloplasts is more common.

In shoots, proplastids from shoot apical meristems can gradually develop into chloroplasts in photosynthetic leaf tissues as the leaf matures, if exposed to the required light. This process involves invaginations of the inner plastid membrane, forming sheets of membrane that project into the internal stroma. These membrane sheets then fold to form thylakoids and grana.

If angiosperm shoots are not exposed to the required light for chloroplast formation, proplastids may develop into an etioplast stage before becoming chloroplasts. An etioplast is a plastid that lacks chlorophyll, and has inner membrane invaginations that form a lattice of tubes in their stroma, called a prolamellar body. While etioplasts lack chlorophyll, they have a yellow chlorophyll precursor stocked. Within a few minutes of light exposure, the prolamellar body begins to reorganize into stacks of thylakoids, and chlorophyll starts to be produced. This process, where the etioplast becomes a chloroplast, takes several hours. Gymnosperms do not require light to form chloroplasts.

Light, however, does not guarantee that a proplastid will develop into a chloroplast. Whether a proplastid develops into a chloroplast some other kind of plastid is mostly controlled by the nucleus and is largely influenced by the kind of cell it resides in.

Plastid Interconversion

Plastid differentiation is not permanent, in fact many interconversions are possible. Chloroplasts may be converted to chromoplasts, which are pigment-filled plastids responsible for the bright colours you see in flowers and ripe fruit. Starch storing amyloplasts can also be converted to chromoplasts, and it is possible for proplastids to develop straight into chromoplasts. Chromoplasts and amyloplasts can also become chloroplasts, like what happens when you illuminate a carrot or a potato. If a plant is injured, or something else causes a plant cell to revert to a meristematic state, chloroplasts and other plastids can turn back into proplastids. Chloroplast, amyloplast, chromoplast, proplast, etc., are not absolute states—intermediate forms are common.

Chloroplast Division

Most chloroplasts in a photosynthetic cell do not develop directly from proplastids or etioplasts. In fact, a typical shoot meristematic plant cell contains only 7–20 proplastids. These proplastids differentiate into chloroplasts, which divide to create the 30–70 chloroplasts found in a mature photosynthetic plant cell. If the cell divides, chloroplast division provides the additional chloroplasts to partition between the two daughter cells.

In single-celled algae, chloroplast division is the only way new chloroplasts are formed. There is no proplastid differentiation—when an algal cell divides, its chloroplast divides along with it, and each daughter cell receives a mature chloroplast.

Almost all chloroplasts in a cell divide, rather than a small group of rapidly dividing chloroplasts. Chloroplasts have no definite S-phase—their DNA replication is not synchronized or limited to that of their host cells. Much of what we know about chloroplast division comes from studying organisms like *Arabidopsis* and the red alga *Cyanidioschyzon merolœ.*

The division process starts when the proteins FtsZ1 and FtsZ2 assemble into filaments, and with the help of a protein ARC6, form a structure called a Z-ring within the chloroplast's stroma. The Min system manages the placement of the Z-ring, ensuring that the chloroplast is cleaved more or less evenly. The protein MinD prevents FtsZ from linking up and forming filaments. Another protein ARC3 may also be involved, but it is not very well understood. These proteins are active at the poles of the chloroplast, preventing Z-ring formation there, but near the centre of the chloroplast, MinE inhibits them, allowing the Z-ring to form.

Next, the two plastid-dividing rings, or PD rings form. The inner plastid-dividing ring is located in the inner side of the chloroplast's inner membrane, and is formed first. The outer plastid-dividing ring is found wrapped around the outer chloroplast membrane. It consists of filaments about 5 nanometers across, arranged in rows 6.4 nanometers apart, and shrinks to squeeze the chloroplast. This is when chloroplast constriction begins.

In a few species like *Cyanidioschyzon merolœ*, chloroplasts have a third plastid-dividing ring located in the chloroplast's intermembrane space.

Late into the constriction phase, dynamin proteins assemble around the outer plastid-dividing ring, helping provide force to squeeze the chloroplast. Meanwhile, the Z-ring and the inner plastid-dividing ring break down. During this stage, the many chloroplast DNA plasmids floating around in the stroma are partitioned and distributed to the two forming daughter chloroplasts.

Later, the dynamins migrate under the outer plastid dividing ring, into direct contact with the chloroplast's outer membrane, to cleave the chloroplast in two daughter chloroplasts.

A remnant of the outer plastid dividing ring remains floating between the two daughter chloroplasts, and a remnant of the dynamin ring remains attached to one of the daughter chloroplasts.

Of the five or six rings involved in chloroplast division, only the outer plastid-dividing ring is present for the entire constriction and division phase—while the Z-ring forms first, constriction does not begin until the outer plastid-dividing ring forms.

Regulation

In species of algae which contain a single chloroplast, regulation of chloroplast division is extremely important to ensure that each daughter cell receives a chloroplast—chloroplasts can't be made from scratch. In organisms like plants, whose cells contain multiple chloroplasts, coordination is looser and less important. It is likely that chloroplast and cell division are somewhat synchronized, though the mechanisms for it are mostly unknown. Light has been shown to be a requirement for chloroplast division. Chloroplasts can grow and progress through some of the constriction stages under poor quality green light, but are slow to complete division—they require exposure to bright white light to complete division. Spinach leaves grown under green light have been observed to contain many large dumbbell-shaped chloroplasts. Exposure to white light can stimulate these chloroplasts to divide and reduce the population of dumbbell-shaped chloroplasts.

Chloroplast Inheritance

Like mitochondria, chloroplasts are usually inherited from a single parent. Biparental chloroplast inheritance—where plastid genes are inherited from both parent plants—occurs in very low levels in some flowering plants.

Many mechanisms prevent biparental chloroplast DNA inheritance including selective destruction of chloroplasts or their genes within the gamete or zygote, and chloroplasts from one parent being excluded from the embryo. Parental chloroplasts can be sorted so that only one type is present in each offspring.

Gymnosperms, such as pine trees, mostly pass on chloroplasts paternally, while flowering plants often inherit chloroplasts maternally. Flowering plants were once thought to only inherit chloroplasts maternally. However, there are now many documented cases of angiosperms inheriting chloroplasts paternally.

Angiosperms which pass on chloroplasts maternally have many ways to prevent paternal inheritance. Most of them produce sperm cells which do not contain any plastids. There are many other documented mechanisms that prevent paternal inheritance in these flowering plants, such as different rates of chloroplast replication within the embryo.

Among angiosperms, paternal chloroplast inheritance is observed more often in hybrids than in offspring from parents of the same species. This suggests that incompatible hybrid genes might interfere with the mechanisms that prevent paternal inheritance.

Transplastomic Plants

Recently, chloroplasts have caught attention by developers of genetically modified crops. Since in most flowering plants, chloroplasts are not inherited from the male parent, transgenes in these plastids cannot be disseminated by pollen. This makes plastid transformation a valuable tool for the creation and cultivation of genetically modified plants that are biologically contained, thus posing significantly lower environmental risks. This biological containment strategy is therefore suitable for establishing the coexistence of conventional and organic agriculture. While the reliability of this mechanism has not yet been studied for all relevant crop species, recent results in tobacco plants are promising, showing a failed containment rate of transplastomic plants at 3 in 1,000,000.

Thylakoid

A thylakoid is a membrane-bound compartment inside chloroplasts and cyanobacteria. They are the site of the light-dependent reactions of photosynthesis. Thylakoids consist of a thylakoid membrane surrounding a thylakoid lumen. Chloroplast thylakoids frequently form stacks of disks referred to as grana (singular: granum). Grana are connected by intergranal or stroma thylakoids, which join granum stacks together as a single functional compartment.

Etymology

The word *thylakoid* come via Latin from Greek *thylakos* meaning "sac" or "pouch". Thus, *thylakoid* means "sac-like" or "pouch-like".

Thylakoid Structure

Thylakoids are membrane-bound structures embedded into the chloroplast stroma. It has a coin shaped structure such that when viewed as a stack of thylakoids in a granum its appearance strongly resembles that of a stack of coins.

Membrane

The thylakoid membrane is the site of the light-dependent reactions of photosynthesis with the photosynthetic pigments embedded directly in the membrane. It is an alternating pattern of dark and light bands measuring each 1 nanometre. The thylakoid lipid bilayer shares characteristic features with prokaryotic membranes and the inner chloroplast membrane. For example, acidic lipids can be found in thylakoid membranes, cyanobacteria and other photosynthetic bacteria and are involved in the functional integrity of the photosystems. The thylakoid membranes of higher plants are composed primarily of phospholipids and galactolipids that are asymmetrically arranged along and across the membranes. The lipids for the thylakoid membranes are synthesized in a complex pathway involving exchange of lipid precursors between the endoplasmic reticulum and inner membrane of the plastid envelope and transported from the inner membrane to the thylakoids via vesicles.

Lumen

The thylakoid lumen is a continuous aqueous phase enclosed by the thylakoid membrane. It plays a vital role for photophosphorylation during photosynthesis. During the light-dependent reaction, protons are pumped across the thylakoid membrane into the lumen making it acidic down to pH 4.

Granum and Stroma Lamellae

In higher plants thylakoids are organized into a granum-stroma membrane assembly. A granum (plural grana) is a stack of thylakoid discs. Chloroplasts can have from 10 to 100 grana. Grana are connected by stroma thylakoids, also called intergrana thylakoids or lamellae. Grana thylakoids and stroma thylakoids can be distinguished by their different protein composition. Grana contribute to chloroplasts' large surface area to volume ratio. Different interpretations of electron tomography imaging of thylakoid membranes has resulted in two models for grana structure. Both posit that lamellae intersect grana stacks in parallel sheets, though whether these sheets intersect in planes perpendicular to the grana stack axis, or are arranged in a right-handed helix is debated.

Thylakoid Formation

Chloroplasts develop from proplastids when seedlings emerge from the ground. Thylakoid formation requires light. In the plant embryo and in the absence of light, proplastids develop into etioplasts that contain semicrystalline membrane structures called prolamellar bodies. When exposed to light, these prolamellar bodies develop into thylakoids. This does not happen in seedlings grown in the dark, which undergo etiolation. An underexposure to light can cause the thylakoids to fail. This causes the chloroplasts to fail resulting in the death of the plant.

Thylakoid formation requires the action of *vesicle-inducing protein in plastids 1* (VIPP1). Plants cannot survive without this protein, and reduced VIPP1 levels lead to slower growth and paler plants with reduced ability to photosynthesize. VIPP1 appears to be required for basic thylakoid membrane formation, but not for the assembly of protein complexes of the thylakoid membrane. It is conserved in all organisms containing thylakoids, including cyanobacteria, green algae, such as Chlamydomonas, and higher plants, such as *Arabidopsis thaliana*.

Thylakoid Isolation and Fractionation

Thylakoids can be purified from plant cells using a combination of differential and gradient centrifugation. Disruption of isolated thylakoids, for example by mechanical shearing, releases the lumenal fraction. Peripheral and integral membrane fractions can be extracted from the remaining membrane fraction. Treatment with sodium carbonate (Na_2CO_3) detaches peripheral membrane proteins, whereas treatment with detergents and organic solvents solubilizes integral membrane proteins.

Thylakoid Proteins

Thylakoids contain many integral and peripheral membrane proteins, as well as lumenal proteins. Recent proteomics studies of thylakoid fractions have provided further details on the protein composition of the thylakoids. These data have been summarized in several plastid protein databases that are available online.

According to these studies, the thylakoid proteome consists of at least 335 different proteins. Out of these, 89 are in the lumen, 116 are integral membrane proteins, 62 are peripheral proteins on the stroma side, and 68 peripheral proteins on the lumenal side. Additional low-abundance lumenal proteins can be predicted through computational methods. Of the thylakoid proteins with known functions, 42% are involved in photosynthesis. The next largest functional groups include proteins involved in protein targeting, processing and folding with 11%, oxidative stress response (9%) and translation (8%).

Integral Membrane Proteins

Thylakoid membranes contain integral membrane proteins which play an important role in light harvesting and the light-dependent reactions of photosynthesis. There are four major protein complexes in the thylakoid membrane:

- Photosystems I and II
- Cytochrome b6f complex
- ATP synthase

Photosystem II is located mostly in the grana thylakoids, whereas photosystem I and ATP synthase are mostly located in the stroma thylakoids and the outer layers of grana. The cytochrome b6f complex is distributed evenly throughout thylakoid membranes. Due to the separate location of the two photosystems in the thylakoid membrane system, mobile electron carriers are required to shuttle electrons between them. These carriers are plastoquinone and plastocyanin. Plastoquinone shuttles electrons from photosystem II to the cytochrome b6f complex, whereas plastocyanin carries electrons from the cytochrome b6f complex to photosystem I.

Together, these proteins make use of light energy to drive electron transport chains that generate a chemiosmotic potential across the thylakoid membrane and NADPH, a product of the terminal redox reaction. The ATP synthase uses the chemiosmotic potential to make ATP during photophosphorylation.

Photosystems

These photosystems are light-driven redox centres, each consisting of an antenna complex that uses chlorophylls and accessory photosynthetic pigments such as carotenoids and phycobiliproteins to harvest light at a variety of wavelengths. Each antenna complex has between 250 and 400 pigment molecules and the energy they absorb is shuttled by resonance energy transfer to a specialized chlorophyll *a* at the reaction centre of each photosystem. When either of the two chlorophyll *a* molecules at the reaction centre absorbs energy, an electron is excited and transferred to an electron-acceptor molecule. Photosystem I contains a pair of chlorophyll *a* molecules, designated P700, at its reaction centre that maximally absorbs 700 nm light. Photosystem II contains P680 chlorophyll that absorbs 680 nm light best (note that these wavelengths correspond to deep red). The P is short for pigment and the number is the specific absorption peak in nanometers for the chlorophyll molecules in each reaction centre.

Cytochrome b6f Complex

The cytochrome b6f complex is part of the thylakoid electron transport chain and couples electron transfer to the pumping of protons into the thylakoid lumen. Energetically, it is situated between the two photosystems and transfers electrons from photosystem II-plastoquinone to plastocyanin-photosystem I.

ATP Synthase

The thylakoid ATP synthase is a CF1FO-ATP synthase similar to the mitochondrial ATPase. It is integrated into the thylakoid membrane with the CF1-part sticking into stroma. Thus, ATP synthesis occurs on the stromal side of the thylakoids where the ATP is needed for the light-independent reactions of photosynthesis.

Thylakoid Lumen Proteins

The electron transport protein plastocyanin is present in the lumen and shuttles electrons from the cytochrome b6f protein complex to photosystem I. While plastoquinones are lipid-soluble and therefore move within the thylakoid membrane, plastocyanin moves through the thylakoid lumen.

The lumen of the thylakoids is also the site of water oxidation by the oxygen evolving complex associated with the lumenal side of photosystem II.

Lumenal proteins can be predicted computationally based on their targeting signals. In Arabidopsis, out of the predicted lumenal proteins possessing the Tat signal, the largest groups with known functions are 19% involved in protein processing (proteolysis and folding), 18% in photosynthesis, 11% in metabolism, and 7% redox carriers and defence.

Thylakoid Protein Expression

Chloroplasts have their own genome, which encodes a number of thylakoid proteins. However, during the course of plastid evolution from their cyanobacterial endosymbiotic ancestors, extensive gene transfer from the chloroplast genome to the cell nucleus took place. This results in the four major thylakoid protein complexes being encoded in part by the chloroplast genome and in part by the nuclear genome. Plants have developed several mechanisms to co-regulate the expression of the different subunits encoded in the two different organelles to assure the proper stoichiometry and assembly of these protein complexes. For example, transcription of nuclear genes encoding parts of the photosynthetic apparatus is regulated by light. Biogenesis, stability and turnover of thylakoid protein complexes are regulated by phosphorylation via redox-sensitive kinases in the thylakoid membranes. The translation rate of chloroplast-encoded proteins is controlled by the presence or absence of assembly partners (control by epistasy of synthesis). This mechanism involves negative feedback through binding of excess protein to the 5' untranslated region of the chloroplast mRNA. Chloroplasts also need to balance the ratios of photosystem I and II for the electron transfer chain. The redox state of the electron carrier plastoquinone in the thylakoid membrane directly affects the transcription of chloroplast genes encoding proteins of the reaction centres of the photosystems, thus counteracting imbalances in the electron transfer chain.

Protein Targeting to the Thylakoids

Thylakoid proteins are targeted to their destination via signal peptides and prokaryotic-type secretory pathways inside the chloroplast. Most thylakoid proteins encoded by a plant's nuclear genome need two targeting signals for proper localization: An N-terminal chloroplast targeting peptide, followed by a thylakoid targeting peptide (shown in blue). Proteins are imported through the translocon of outer and inner membrane (Toc and Tic) complexes. After entering the chloroplast, the first targeting peptide is cleaved off by a protease processing imported proteins. This unmasks the second targeting signal and the protein is exported from the stroma into the thylakoid in a second targeting step. This second step requires the action of protein translocation components of the thylakoids and is energy-dependent. Proteins are inserted into the membrane via the SRP-dependent pathway (1), the Tat-dependent pathway (2), or spontaneously via their transmembrane domains. Lumenal proteins are exported across the thylakoid membrane into the lumen by either the Tat-dependent pathway (2) or the Sec-dependent pathway (3) and released by cleavage from the thylakoid targeting signal. The different pathways utilize different signals and energy sources. The Sec (secretory) pathway requires ATP as energy source and consists of SecA, which binds to the imported protein and a Sec membrane complex to shuttle the protein across. Proteins with a twin arginine motif in their thylakoid signal peptide are shuttled through the Tat (twin arginine translocation) pathway, which requires a membrane-bound Tat complex and the pH gradient as an energy source. Some other proteins are inserted into the membrane via the SRP (signal recognition particle) pathway. The chloroplast SRP can interact with its target proteins either post-translationally or co-translationally, thus transporting imported proteins as well as those that are translated inside the chloroplast. The SRP pathway requires GTP and the pH gradient as energy sources. Some transmembrane proteins may also spontaneously insert into the membrane from the stromal side without energy requirement.

Thylakoid Function

The thylakoids are the site of the light-dependent reactions of photosynthesis. These include light-driven water oxidation and oxygen evolution, the pumping of protons across the thylakoid membranes coupled with the electron transport chain of the photosystems and cytochrome b6f complex, and ATP synthesis by the ATP synthase utilizing the generated proton gradient.

Water Photolysis

The first step in photosynthesis is the light-driven reduction (splitting) of water to provide the electrons for the photosynthetic electron transport chains as well as protons for the establishment of a proton gradient. The water-splitting reaction occurs on the lumenal side of the thylakoid membrane and is driven by the light energy captured by the photosystems. It is interesting to note that this reduction of water conveniently produces the waste product O_2 that is vital for cellular respiration. The molecular oxygen formed by the reaction is released into the atmosphere.

Electron Transport Chains

Two different variations of electron transport are used during photosynthesis:

- Noncyclic electron transport or Non-cyclic photophosphorylation produces NADPH + H^+ and ATP.
- Cyclic electron transport or Cyclic photophosphorylation produces only ATP.

The noncyclic variety involves the participation of both photosystems, while the cyclic electron flow is dependent on only photosystem I.

- Photosystem I uses light energy to reduce $NADP^+$ to NADPH + H^+, and is active in both noncyclic and cyclic electron transport. In cyclic mode, the energized electron is passed down a chain that ultimately returns it (in its base state) to the chlorophyll that energized it.
- Photosystem II uses light energy to oxidize water molecules, producing electrons (e^-), protons (H^+), and molecular oxygen (O_2), and is only active in noncyclic transport. Electrons in this system are not conserved, but are rather continually entering from oxidized $2H_2O$ (O_2 + 4 H^+ + 4 e^-) and exiting with $NADP^+$ when it is finally reduced to NADPH.

Chemiosmosis

A major function of the thylakoid membrane and its integral photosystems is the establishment of chemiosmotic potential. The carriers in the electron transport chain use some of the electron's energy to actively transport protons from the stroma to the lumen. During photosynthesis, the lumen becomes acidic, as low as pH 4, compared to pH 8 in the stroma. This represents a 10,000 fold concentration gradient for protons across the thylakoid membrane.

Source of Proton Gradient

The protons in the lumen come from three primary sources.

- Photolysis by photosystem II oxidises water to oxygen, protons and electrons in the lumen.
- The transfer of electrons from photosystem II to plastoquinone during non-cyclic electron transport consumes two protons from the stroma. These are released in the lumen when the reduced plastoquinol is oxidized by the cytochrome b6f protein complex on the lumen side of the thylakoid membrane. From the plastoquinone pool, electrons pass through the cytochrome b6f complex. This integral membrane assembly resembles cytochrome bc1.
- The reduction of plastoquinone by ferredoxin during cyclic electron transport also transfers two protons from the stroma to the lumen.

The proton gradient is also caused by the consumptions of protons in the stroma to make NADPH from NADP+ at the NADP reductase.

ATP Generation

The molecular mechanism of ATP generation in chloroplasts is similar to that in mitochondria and takes the required energy from the proton motive force (PMF). However, chloroplasts rely more on the chemical potential of the PMF to generate the potential energy required for ATP synthesis. The PMF is the sum of a proton chemical potential (given by the proton concentration gradient) and a transmembrane electrical potential (given by charge separation across the membrane). Compared to the inner membranes of mitochondria, which have a significantly higher membrane potential due to charge separation, thylakoid membranes lack a charge gradient. To compensate for this, the 10,000 fold proton concentration gradient across the thylakoid membrane is much higher compared to a 10 fold gradient across the inner membrane of mitochondria. The resulting chemiosmotic potential between the lumen and stroma is high enough to drive ATP synthesis using the ATP synthase. As the protons travel back down the gradient through channels in ATP synthase, ADP + P_i are combined into ATP. In this manner, the light-dependent reactions are coupled to the synthesis of ATP via the proton gradient.

Thylakoid Membranes in Cyanobacteria

Cyanobacteria are photosynthetic prokaryotes with highly differentiated membrane systems. Cyanobacteria have an internal system of thylakoid membranes where the fully functional electron transfer chains of photosynthesis and respiration reside. The presence of different membrane systems lends these cells a unique complexity among bacteria. Cyanobacteria must be able to reorganize the membranes, synthesize new membrane lipids, and properly target proteins to the correct membrane system. The outer membrane, plasma membrane, and thylakoid membranes each have specialized roles in the cyanobacterial cell. Understanding the organization, functionality, protein composition and dynamics of the membrane systems remains a great challenge in cyanobacterial cell biology.

The thylakoid membranes of the cyanobacteria are not differentiated into granum and stroma regions as it happens in plants. They form stacks of parallel sheets close to the cytoplasmic membrane with a low packing density. The relatively large distance between the thylakoids provides space for the external light harvesting antennae, the phycobilisomes. This macrostructure, as in the case of higher plants, shows some flexibility during changes in the physicochemical environment.

Light-dependent Reactions

The light-dependent reactions, or photoreduction, is the first stage of photosynthesis, is a process by which plants capture and store energy from sunlight. In this process, light energy is converted into chemical energy, in the form of the energy-carrying molecules ATP and NADPH. In the *light-independent reactions*, the formed NADPH and ATP drive the reduction of CO

2 to more useful organic compounds, such as glucose. However, although light-independent reactions are, by convention, also called *dark reactions*, they are not independent of the need of light, for they are driven by ATP and NADPH, products of light. They are often called the Calvin Cycle or C3 Cycle.

The light-dependent reactions take place on the thylakoid membranes. The inside of the thylakoid membrane is called the lumen, and outside the thylakoid membrane is the stroma, where the light-independent reactions take place. The thylakoid membrane contains some integral membrane protein complexes that catalyze the light reactions. There are four major protein complexes in the thylakoid membrane: Photosystem II (PSII), Cytochrome b6f complex, Photosystem I (PSI), and ATP synthase. These four complexes work together to ultimately create the products ATP and NADPH.

The two photosystems absorb light energy through pigments - primarily the chlorophylls, which are responsible for the green colour of leaves. The light-dependent reactions begin in photosystem II. When a chlorophyll *a* molecule within the reaction centre of PSII absorbs a photon, an electron in this molecule attains a higher energy level. Because this state of an electron is very unstable, the electron is transferred from one to another molecule creating a chain of redox reactions, called an electron transport chain (Etc.). The electron flow goes from PSII to cytochrome b6f to PSI. In PSI, the electron gets the energy from another photon. The final electron acceptor is NADP. In *oxygenic photosynthesis*, the first electron donor is water, creating oxygen as a waste product. In *anoxygenic photosynthesis* various electron donors are used.

Cytochrome b6f and ATP synthase work together to create ATP. This process is called photophosphorylation, which occurs in two different ways. In *non-cyclic photophosphorylation*, cytochrome b6f uses the energy of electrons from PSII to pump protons from the stroma to the lumen. The proton gradient across the thylakoid membrane creates a proton-motive force, used by ATP synthase to form ATP. In *cyclic photophosphorylation*, cytochrome b6f uses the energy of electrons from not only PSII but also PSI to create more ATP and to stop the production of NADPH. Cyclic phosphorylation is important to create ATP and maintain NADPH in the right proportion for the light-independent reactions.

The net-reaction of all light-dependent reactions in oxygenic photosynthesis is:

$$2H_2O + 2NADP{+}{+}\ 3ADP + 3P_i \rightarrow O_2 + 2NADPH + 3ATP$$

The two photosystems are protein complexes that absorb photons and are able to use this energy to create an electron transport chain. Photosystem I and II are very similar in structure and function. They use special proteins, called light-harvesting complexes, to absorb the photons with very high effectiveness. If a special pigment molecule in a photosynthetic reaction centre absorbs a photon, an electron in this pigment attains the excited state and then is transferred to another molecule in the reaction centre. This reaction, called *photoinduced charge separation*, is the start of the electron flow and is unique because it transforms light energy into chemical forms.

The Reaction Centre

The reaction centre is in the thylakoid membrane. It transfers light energy to a dimer of chlorophyll pigment molecules near the periplasmic (or thylakoid lumen) side of the membrane. This dimer is called a special pair because of its fundamental role in photosynthesis. This special pair is slightly different in PSI and PSII reaction centre. In PSII, it absorbs photons with a wavelength of 680 nm, and it is therefore called P680. In PSI, it absorbs photons at 700 nm, and it is called P700. In bacteria, the special pair is called P760, P840, P870, or P960.

If an electron of the special pair in the reaction centre becomes excited, it cannot transfer this energy to another pigment using resonance energy transfer. In normal circumstances, the electron should return to the ground state, but, because the reaction centre is arranged so that a suitable electron acceptor is nearby, the excited electron can move from the initial molecule to the acceptor. This process results in the formation of a positive charge on the special pair (due to the loss of an electron) and a negative charge on the acceptor and is, hence, referred to as photoinduced charge separation. In other words, electrons in pigment molecules can exist at specific energy levels. Under normal circumstances, they exist at the lowest possible energy level they can. However, if there is enough energy to move them into the next energy level, they can absorb that energy and occupy that higher energy level. The light they absorb contains the necessary amount of energy needed to push them into the next level. Any light that does not have enough or has too much energy cannot be absorbed and is reflected. The electron in the higher energy level, however, does not want to be there; the electron is unstable and must return to its normal lower energy level. To do this, it must release the energy that has put it into the higher energy state to begin with. This can happen various ways. The extra energy can be converted into molecular motion and lost as heat. Some of the extra energy can be lost as heat energy, while the rest is lost as light. This re-emission of light energy is called fluorescence. The energy, but not the e- itself, can be passed onto another molecule. This is called resonance. The energy and the e- can be transferred to another molecule. Plant pigments usually utilize the last two of these reactions to convert the sun's energy into their own.

This initial charge separation occurs in less than 10 picoseconds (10^{-11} seconds). In their high-energy states, the special pigment and the acceptor could undergo charge recombination; that is, the electron on the acceptor could move back to neutralize the positive charge on the special pair. Its return to the special pair would waste a valuable high-energy electron and simply convert the absorbed light energy into heat. In the case of PSII, this backflow of electrons can produce reactive oxygen species leading to photoinhibition. Three factors in the structure of the reaction centre work together to suppress charge recombination nearly completely.

- Another electron acceptor is less than 10 Å away from the first acceptor, and so the electron is rapidly transferred farther away from the special pair.
- An electron donor is less than 10 Å away from the special pair, and so the positive charge is neutralized by the transfer of another electron
- The electron transfer back from the electron acceptor to the positively charged special pair is especially slow. The rate of an of electron transfer reaction increases with its thermodynamic favourability up to a point and then decreases. The back transfer is so favourable that it takes place in the inverted region where electron-transfer rates become slower.

Thus, electron transfer proceeds efficiently from the first electron acceptor to the next, creating an electron transport chain that ends if it has reached NADPH.

Photosynthetic Electron Transport Chains in Chloroplasts

The photosynthesis process in chloroplasts begins when an electron of P680 of PSII attains an higher-energy level. This energy is used to reduce a chain of electron acceptors that have subsequently lowered redox-potentials. This chain of electron acceptors is known as an electron transport chain. When this chain reaches PS I, an electron is again excited, creating a high redox-potential. The electron transport chain of photosynthesis is often put in a diagram called the z-scheme, because the redox diagram from P680 to P700 resembles the letter z.

The final product of PSII is plastoquinol, a mobile electron carrier in the membrane. Plastoquinol transfers the electron from PSII to the proton pump, cytochrome b6f. The ultimate electron donor of PSII is water. Cytochrome b6f proceeds the electron chain to PSI through plastocyanin molecules. PSI is able to continue the electron transfer in two different ways. It can transfer the electrons either to plastoquinol again, creating a cyclic electron flow, or to an enzyme called FNR (Ferredoxin—NADP(+) reductase), creating a non-cyclic electron flow. PSI releases FNR into the stroma, where it reduces NADP+ to NADPH.

Activities of the electron transport chain, especially from cytochrome b6f, lead to pumping of protons from the stroma to the lumen. The resulting transmembrane proton gradient is used to make ATP via ATP synthase.

The overall process of the photosynthetic electron transport chain in chloroplasts is:

H_2O → PS II → plastoquinone → *cytb6* → plastocyanin → PS I → NADPH

Photosystem II

PS II is an extremely complex, highly organized transmembrane structure that contains a *water-splitting complex*, chlorophylls and carotenoid pigments, a *reaction centre* (P680), pheophytin (a pigment similar to chlorophyll), and two quinones. It uses the energy of sunlight to transfer electrons from water to a mobile electron carrier in the membrane called *plastoquinone*:

H_2O → P680 → P680* → plastoquinone

Plastoquinone, in turn, transfers electrons to *cytb*

6, which feeds them into PS I.

The Water-splitting Complex

The step H_2O → P680 is performed by a poorly understood structure embedded within PS II called the *water-splitting complex* or the *oxygen-evolving complex*. It catalyzes a reaction that splits water into electrons, protons and oxygen:

$2H_2O \rightarrow 4H^+ + 4e^- + O_2$

The electrons are transferred to special chlorophyll molecules (embedded in PS II) that are promoted to a higher-energy state by the energy of photons.

The Reaction Centre

The excitation P680 → P680*of the reaction centre pigment P680 occurs here. These special chlorophyll molecules embedded in PS II absorb the energy of photons, with maximal absorption at 680 nm. Electrons within these molecules are promoted to a higher-energy state. This is one of two core processes in photosynthesis, and it occurs with astonishing efficiency (greater than 90%) because, in addition to direct excitation by light at 680 nm, the energy of light first harvested by *antenna proteins* at other wavelengths in the light-harvesting system is also transferred to these special chlorophyll molecules.

This is followed by the step P680→ pheophytin, and then on to plastoquinone, which occurs within the reaction centre of PS II. High-energy electrons are transferred to plastoquinone before it subsequently picks up two protons to become plastoquinol. Plastoquinol is then released into the membrane as a mobile electron carrier.

This is the second core process in photosynthesis. The initial stages occur within *picoseconds*, with an efficiency of 100%. The seemingly impossible efficiency is due to the precise positioning of molecules within the reaction centre. This is a solid-state process, not a chemical reaction. It occurs within an essentially crystalline environment created by the macromolecular structure of PS II. The usual rules of chemistry (which involve random collisions and random energy distributions) do not apply in solid-state environments.

Link of Water-splitting Complex and Chlorophyll Excitation

When the chlorophyll passes the electron to pheophytin, it obtains an electron from P_{680}^*. In turn, P_{680}^* can oxidize the Z (or Y_Z) molecule. Once oxidized, the Z molecule can derive electrons from the water-splitting complex.

Summary

PS II is a transmembrane structure found in all chloroplasts. It splits water into electrons, protons and molecular oxygen. The electrons are transferred to plastoquinone, which carries them to a proton pump. Molecular oxygen is released into the atmosphere.

The emergence of such an incredibly complex structure, a macromolecule that converts the energy of sunlight into potentially useful work with efficiencies that are impossible in ordinary experience, seems almost magical at first glance. Thus, it is of considerable interest that, in essence, the same structure is found in *purple bacteria.*

Cytochrome b6

PS II and PS I are connected by a transmembrane proton pump, cytochrome *b*

6 complex (plastoquinol—plastocyanin reductase; EC 1.10.99.1). Electrons from PS II are carried by plastoquinol to cyt*b*

6, where they are removed in a stepwise fashion (reforming plastoquinone) and transferred to a water-soluble electron carrier called *plastocyanin.* This redox process is coupled to the pumping of four protons across the membrane. The resulting proton gradient (together with the proton gradient produced by the water-splitting complex in PS II) is used to make ATP via ATP synthase.

The similarity in structure and function between cytochrome *b*

6 (in chloroplasts) and cytochrome *bc*

1 (*Complex III* in mitochondria) is striking. Both are transmembrane structures that remove electrons from a mobile, lipid-soluble electron carrier (plastoquinone in chloroplasts; ubiquinone in mitochondria) and transfer them to a mobile, water-soluble electron carrier (plastocyanin in chloroplasts; cytochrome *c* in mitochondria). Both are proton pumps that produce a transmembrane proton gradient.

History

The first ideas about light's being used in photosynthesis were proposed by Colin Flannery in 1779 who recognized it was sunlight falling on plants that was required, although Joseph Priestley had noted the production of oxygen without the association with light in 1772. Cornelius Van Niel proposed in 1931 that photosynthesis is a case of general mechanism where a photon of light is used to photo decompose a hydrogen donor and the hydrogen being used to reduce CO

2. Then in 1939 Robin Hill showed that isolated chloroplasts would make oxygen, but not fix CO

2 showing the light and dark reactions occurred in different places. This led later to the discovery of photosystem 1 and 2.

Z scheme

In plants, light-dependent reactions occur in the thylakoid membranes of the chloroplasts and use light energy to synthesize ATP and NADPH. The light-dependent reaction has two forms: cyclic and non-cyclic. In the non-cyclic reaction, the photons are captured in the light-harvesting antenna complexes of photosystem II by chlorophyll and other accessory pigments. When a chlorophyll molecule at the core of the photosystem II reaction centre obtains sufficient excitation energy from the adjacent antenna pigments, an electron is transferred to the primary electron-acceptor molecule, pheophytin, through a process called photoinduced charge separation. These electrons are shuttled through an electron transport chain, the so-called *Z-scheme* shown in the diagram, that initially functions to generate a chemiosmotic potential across the membrane. An ATP synthase enzyme uses the chemiosmotic potential to make ATP during photophosphorylation, whereas NADPH is a product of the terminal redox reaction in the *Z-scheme*. The electron enters a chlorophyll molecule in Photosystem I. The electron is excited due to the light absorbed by the photosystem. A second electron carrier accepts the electron, which again is passed down lowering energies of electron acceptors. The energy created by the electron acceptors is used to move hydrogen ions across the thylakoid membrane into the lumen. The electron is used to reduce the co-enzyme NADP, which has functions in the light-independent reaction. The cyclic reaction is similar to that of the non-cyclic, but differs in the form that it generates only ATP, and no reduced NADP (NADPH) is created. The cyclic reaction takes place only at photosystem I. Once the electron is displaced from the photosystem, the electron is passed down the electron acceptor molecules and returns to photosystem I, from where it was emitted, hence the name *cyclic reaction.*

Water Photolysis

The NADPH is the main reducing agent in chloroplasts, providing a source of energetic electrons to other reactions. Its production leaves chlorophyll with a deficit of electrons (oxidized), which must be obtained from some other reducing agent. The excited electrons lost from chlorophyll in photosystem I are replaced from the electron transport chain by plastocyanin. However, since photosystem II includes the first steps of the *Z-scheme*, an external source of electrons is required to reduce its oxidized chlorophyll *a* molecules. The source of electrons in green-plant and cyanobacterial photosynthesis is water. Two water molecules are oxidized by four successive charge-separation reactions by photosystem II to yield a molecule of diatomic oxygen and four hydrogen ions; the electron yielded in each step is transferred to a redox-active tyrosine residue that then reduces the photoxidized paired-chlorophyll *a* species called P680 that serves as the primary (light-driven) electron donor in the photosystem II reaction centre. The oxidation of water is catalyzed in photosystem II by a redox-active structure that contains four manganese ions and a calcium ion; this oxygen-evolving complex binds two water molecules and stores the four oxidizing equivalents that are required to drive the water-oxidizing reaction. Photosystem II is the only known biological enzyme that carries out this oxidation of water. The hydrogen ions contribute to the transmembrane chemiosmotic potential that leads to ATP synthesis. Oxygen is a waste product of light-dependent reactions, but the majority of organisms on Earth use oxygen for cellular respiration, including photosynthetic organisms.

Light-independent Reactions

Calvin Cycle

In the light-independent (or "dark") reactions, the enzyme RuBisCO captures CO_2 from the atmosphere and in a process that requires the newly formed NADPH, called the Calvin-Benson Cycle, releases three-carbon sugars, which are later combined to form sucrose and starch. The overall equation for the light-independent reactions in green plants is:

$3\ CO_2 + 9\ ATP + 6\ NADPH + 6\ H^+ \rightarrow C_3H_6O_3\text{-phosphate} + 9\ ADP + 8\ P_i + 6\ NADP^+ + 3\ H_2O$

To be more specific, carbon fixation produces an intermediate product, which is then converted to the final carbohydrate products. The carbon skeletons produced by photosynthesis are then variously used to form other organic compounds, such as the building material cellulose, as precursors for lipid and amino acid biosynthesis, or as a fuel in cellular respiration. The latter occurs not only in plants but also in animals when the energy from plants gets passed through a food chain.

The fixation or reduction of carbon dioxide is a process in which carbon dioxide combines with a five-carbon sugar, ribulose 1,5-bisphosphate (RuBP), to yield two molecules of a three-carbon compound, glycerate 3-phosphate (GP), also known as 3-phosphoglycerate (PGA). GP, in the presence of ATP and NADPH from the light-dependent stages, is reduced to glyceraldehyde 3-phosphate (G3P). This product is also referred to as 3-phosphoglyceraldehyde (PGAL) or even as triose phosphate. Triose is a 3-carbon sugar. Most (5 out of 6 molecules) of the G3P produced is used to regenerate RuBP so the process can continue. The 1 out of 6 molecules of the triose phosphates not "recycled" often condense to form hexose phosphates, which ultimately yield sucrose, starch and cellulose. The sugars produced during carbon metabolism yield carbon skeletons that can be used for other metabolic reactions like the production of amino acids and lipids.

Carbon Concentrating Mechanisms

On Land: In hot and dry conditions, plants close their stomata to prevent the loss of water. Under these conditions, CO_2 will decrease, and oxygen gas, produced by the light reactions of photosynthesis, will decrease in the stem, not leaves, causing an increase of photorespiration by the oxygenase activity of ribulose-1,5-bisphosphate carboxylase/ oxygenase and decrease in carbon fixation. Some plants have evolved mechanisms to increase the CO_2 concentration in the leaves under these conditions.

C_4 plants chemically fix carbon dioxide in the cells of the mesophyll by adding it to the three-carbon molecule phosphoenolpyruvate (PEP), a reaction catalyzed by an enzyme called PEP carboxylase, creating the four-carbon organic acid oxaloacetic acid. Oxaloacetic acid or malate synthesized by this process is then translocated to specialized bundle sheath cells where the enzyme RuBisCO and other Calvin cycle enzymes are located, and where CO_2 released by decarboxylation of the four-carbon acids is then fixed by RuBisCO activity to the three-carbon sugar 3-phosphoglyceric acids. The physical separation of RuBisCO from the oxygen-generating light reactions reduces photorespiration and increases CO_2 fixation and, thus, photosynthetic capacity of the leaf. C_4 plants can produce more sugar than C_3 plants in conditions of high light and temperature. Many important crop plants are C_4 plants, including maize, sorghum, sugarcane, and millet. Plants that do not use PEP-carboxylase in carbon fixation are called C_3 plants because the primary carboxylation reaction, catalyzed by RuBisCO, produces the three-carbon sugar 3-phosphoglyceric acids directly in the Calvin-Benson cycle. Over 90% of plants use C_3 carbon fixation, compared to 3% that use C_4 carbon fixation.; however, the fact that C_4 has evolved in over 60 plant lineages makes it a striking example of convergent evolution.

Xerophytes, such as cacti and most succulents, also use PEP carboxylase to capture carbon dioxide in a process called Crassulacean acid metabolism (CAM). In contrast to C_4 metabolism, which *physically* separates the CO_2 fixation to PEP from the Calvin cycle, CAM *temporally* separates these two processes. CAM plants have a different leaf anatomy from C_3 plants, and fix the CO_2 at night, when their stomata are open. CAM plants store the CO_2 mostly in the form of malic acid via carboxylation of phosphoenolpyruvate to oxaloacetate, which is then reduced to malate. Decarboxylation of malate during the day releases CO_2 inside the leaves, thus allowing carbon fixation to 3-phosphoglycerate by RuBisCO. Sixteen thousand species of plants use CAM.

In Water

Cyanobacteria possess carboxysomes, which increase the concentration of CO_2 around RuBisCO to increase the rate of photosynthesis. An enzyme, carbonic anhydrase, located within the carboxysome releases CO_2 from the dissolved hydrocarbonate ions (HCO_3^-). Before the CO_2 diffuses out it is quickly sponged up by RuBisCO, which is concentrated within the carboxysomes. HCO_3^- ions are made from CO_2 outside the cell by another carbonic anhydrase and are actively pumped into the cell by a membrane protein. They cannot cross the membrane as they are charged, and within the cytosol they turn back into CO_2 very slowly without the help of carbonic anhydrase. This causes the HCO_3^- ions to accumulate within the cell from where they diffuse into the carboxysomes. Pyrenoids in algae and hornworts also act to concentrate CO_2 around rubisco.

Order and Kinetics

The overall process of photosynthesis takes place in four stages:

Stage	Description	Time scale
1	Energy transfer in antenna chlorophyll (thylakoid membranes)	femtosecond to picosecond
2	Transfer of electrons in photochemical reactions (thylakoid membranes)	picosecond to nanosecond
3	Electron transport chain and ATP synthesis (thylakoid membranes)	microsecond to millisecond
4	Carbon fixation and export of stable products	millisecond to second

Efficiency

Plants usually convert light into chemical energy with a photosynthetic efficiency of 3–6%. Absorbed light that is unconverted is dissipated primarily as heat, with a small fraction (1-2%) re-emitted as chlorophyll fluorescence at longer (redder) wavelengths.

Actual plants' photosynthetic efficiency varies with the frequency of the light being converted, light intensity, temperature and proportion of carbon dioxide in the atmosphere, and can vary from 0.1% to 8%. By comparison, solar panels convert light into electric energy at an efficiency of approximately 6–20% for mass-produced panels, and above 40% in laboratory devices.

Photosynthesis measurement systems are not designed to directly measure the amount of light absorbed by the leaf. Nevertheless, the light response curves that systems like the LCpro-SD produce, do allow comparisons in photosynthetic efficiency between plants.

Evolution

Early photosynthetic systems, such as those from green and purple sulphur and green and purple nonsulphur bacteria, are thought to have been anoxygenic, using various molecules as electron donors. Green and purple sulphur bacteria are thought to have used hydrogen and sulphur as an electron donor. Green nonsulphur bacteria used various amino and other organic acids. Purple nonsulphur bacteria used a variety of nonspecific organic molecules. The use of these molecules is consistent with the geological evidence that the atmosphere was highly reduced at that time.

Fossils of what are thought to be filamentous photosynthetic organisms have been dated at 3.4 billion years old.

The main source of oxygen in the atmosphere is oxygenic photosynthesis, and its first appearance is sometimes referred to as the oxygen catastrophe. Geological evidence suggests that oxygenic photosynthesis, such as that in cyanobacteria, became important during the Paleoproterozoic era around 2 billion years ago. Modern photosynthesis in plants and most photosynthetic prokaryotes is oxygenic. Oxygenic photosynthesis uses water as an electron donor, which is oxidized to molecular oxygen (O_2) in the photosynthetic reaction centre.

Symbiosis and the Origin of Chloroplasts

Several groups of animals have formed symbiotic relationships with photosynthetic algae. These are most common in corals, sponges and sea anemones. It is presumed that this is due to the particularly simple body plans and large surface areas of these animals compared to their volumes. In addition, a few marine mollusks *Elysia viridis* and *Elysia chlorotica* also maintain a symbiotic relationship with chloroplasts they capture from the algae in their diet and then store in their bodies. This allows the mollusks to survive solely by photosynthesis for several months at a time. Some of the genes from the plant cell nucleus have even been transferred to the slugs, so that the chloroplasts can be supplied with proteins that they need to survive.

An even closer form of symbiosis may explain the origin of chloroplasts. Chloroplasts have many similarities with photosynthetic bacteria, including a circular chromosome, prokaryotic-type ribosomes, and similar proteins in the photosynthetic reaction centre. The endosymbiotic theory suggests that photosynthetic bacteria were acquired (by endocytosis) by early eukaryotic cells to form the first plant cells. Therefore, chloroplasts may be photosynthetic bacteria that adapted to life inside plant cells. Like mitochondria, chloroplasts still possess their own DNA, separate from the nuclear DNA of their plant host cells and the genes in this chloroplast DNA resemble those in cyanobacteria. DNA in chloroplasts codes for redox proteins such as photosynthetic reaction centres. The CoRR Hypothesis proposes that this Co-location is required for Redox Regulation.

Cyanobacteria and the Evolution of Photosynthesis

The biochemical capacity to use water as the source for electrons in photosynthesis evolved once, in a common ancestor of extant cyanobacteria. The geological record indicates that this transforming event took place early in Earth's history, at least 2450–2320 million years ago (Ma), and, it is speculated, much earlier. Available evidence from geobiological studies of Archean (>2500 Ma) sedimentary rocks indicates that life existed 3500 Ma, but the question of when oxygenic photosynthesis evolved is still unanswered. A clear paleontological window on cyanobacterial evolution opened about 2000 Ma, revealing an already-diverse biota of blue-greens. Cyanobacteria remained principal primary producers throughout the Proterozoic Eon (2500–543 Ma), in part because the redox structure of the oceans favoured photoautotrophs capable of nitrogen fixation. Green algae joined blue-greens as major primary producers on continental shelves near the end of the Proterozoic, but only with the Mesozoic (251–65 Ma) radiations of dinoflagellates, coccolithophorids, and diatoms did primary production in marine shelf waters take modern form. Cyanobacteria remain critical to marine ecosystems as primary producers in oceanic gyres, as agents of biological nitrogen fixation, and, in modified form, as the plastids of marine algae.

A 2010 study by researchers at Tel Aviv University discovered that the Oriental hornet (*Vespa orientalis*) converts sunlight into electric power using a pigment called xanthopterin. This is the first scientific evidence of a member of the animal kingdom engaging in photosynthesis.

Discovery

Although some of the steps in photosynthesis are still not completely understood, the overall photosynthetic equation has been known since the 19th century.

Jan van Helmont began the research of the process in the mid-17th century when he carefully measured the mass of the soil used by a plant and the mass of the plant as it grew. After noticing that the soil mass changed very little, he hypothesized that the mass of the growing plant must come from the water, the only substance he added to the potted plant. His hypothesis was partially accurate — much of the gained mass also comes from carbon dioxide as well as water. However, this was a signalling point to the idea that the bulk of a plant's biomass comes from the inputs of photosynthesis, not the soil itself.

Joseph Priestley, a chemist and minister, discovered that, when he isolated a volume of air under an inverted jar, and burned a candle in it, the candle would burn out very quickly, much before it ran out of wax. He further discovered that a mouse could similarly "injure" air. He then showed that the air that had been "injured" by the candle and the mouse could be restored by a plant.

In 1778, Jan Ingenhousz, court physician to the Austrian Empress, repeated Priestley's experiments. He discovered that it was the influence of sunlight on the plant that could cause it to revive a mouse in a matter of hours.

In 1796, Jean Senebier, a Swiss pastor, botanist, and naturalist, demonstrated that green plants consume carbon dioxide and release oxygen under the influence of light. Soon afterward, Nicolas-Théodore de Saussure showed that the increase in mass of the plant as it grows could not be due only to uptake of CO_2 but also to the incorporation of water. Thus, the basic reaction by which photosynthesis is used to produce food (such as glucose) was outlined.

Cornelis Van Niel made key discoveries explaining the chemistry of photosynthesis. By studying purple sulphur bacteria and green bacteria he was the first scientist to demonstrate that photosynthesis is a light-dependent redox reaction, in which hydrogen reduces carbon dioxide.

Robert Emerson discovered two light reactions by testing plant productivity using different wavelengths of light. With the red alone, the light reactions were suppressed. When blue and red were combined, the output was much more substantial. Thus, there were two photosystems, one absorbing up to 600 nm wavelengths, the other up to 700 nm. The former is known as PSII, the latter is PSI. PSI contains only chlorophyll a, PSII contains primarily chlorophyll a with most of the available chlorophyll b, among other pigment. These include phycobilins, which are the red and blue pigments of red and blue algae respectively, and fucoxanthol for brown algae and diatoms. The process is most productive when absorption of quanta are equal in both the PSII and PSI, assuring that input energy from the antenna complex is divided between the PSI and PSII system, which in turn powers the photochemistry.

Robert Hill thought that a complex of reactions consisting of an intermediate to cytochrome b_6 (now a plastoquinone), another is from cytochrome f to a step in the carbohydrate-generating mechanisms. These are linked by plastoquinone, which does require energy to reduce cytochrome f for it is a sufficient reductant. Further experiments to prove that the oxygen developed during the photosynthesis of green plants came from water, were performed by Hill in 1937 and 1939. He showed that isolated chloroplasts give off oxygen in the presence of unnatural reducing agents like iron oxalate, ferricyanide or benzoquinone after exposure to light. The Hill reaction is as follows:

$$2\ H_2O + 2\ A + (\text{light, chloroplasts}) \rightarrow 2\ AH_2 + O_2$$

where A is the electron acceptor. Therefore, in light, the electron acceptor is reduced and oxygen is evolved.

Samuel Ruben and Martin Kamen used radioactive isotopes to determine that the oxygen liberated in photosynthesis came from the water.

Melvin Calvin and Andrew Benson, along with James Bassham, elucidated the path of carbon assimilation (the photosynthetic carbon reduction cycle) in plants. The carbon reduction cycle is known as the Calvin cycle, which ignores the contribution of Bassham and Benson. Many scientists refer to the cycle as the Calvin-Benson Cycle, Benson-Calvin, and some even call it the Calvin-Benson-Bassham (or CBB) Cycle.

Nobel Prize-winning scientist Rudolph A. Marcus was able to discover the function and significance of the electron transport chain.

Otto Heinrich Warburg and Dean Burk discovered the I-quantum photosynthesis reaction that splits the CO_2, activated by the respiration.

Louis N.M. Duysens and Jan Amesz discovered that chlorophyll a will absorb one light, oxidize cytochrome f, chlorophyll a (and other pigments) will absorb another light, but will reduce this same oxidized cytochrome, stating the two light reactions are in series.

Development of the Concept

In 1893, Charles Reid Barnes proposed two terms, *photosyntax* and *photosynthesis*, for the biological process of *synthesis of complex carbon compounds out of carbonic acid, in the presence of chlorophyll, under the influence of light.* Over time, the term *photosynthesis* came into common usage as the term of choice. Later discovery of anoxygenic photosynthetic bacteria and photophosphorylation necessitated redefinition of the term.

Factors

There are three main factors affecting photosynthesis and several corollary factors. The three main are:

- Light irradiance and wavelength
- Carbon dioxide concentration
- Temperature.

Light Intensity (Irradiance), Wavelength and Temperature

In the early 20th century, Frederick Blackman and Gabrielle Matthaei investigated the effects of light intensity (irradiance) and temperature on the rate of carbon assimilation.

- At constant temperature, the rate of carbon assimilation varies with irradiance, initially increasing as the irradiance increases. However, at higher irradiance, this relationship no longer holds and the rate of carbon assimilation reaches a plateau.
- At constant irradiance, the rate of carbon assimilation increases as the temperature is increased over a limited range. This effect is seen only at high irradiance levels. At low irradiance, increasing the temperature has little influence on the rate of carbon assimilation.

These two experiments illustrate vital points: First, from research it is known that, in general, photochemical reactions are not affected by temperature. However, these experiments clearly show that temperature affects the rate of carbon assimilation, so there must be two sets of reactions in the full process of carbon assimilation. These are, of course, the light-dependent 'photochemical' stage and the light-independent, temperature-dependent stage. Second, Blackman's experiments illustrate the concept of limiting factors. Another limiting factor is the wavelength of light. Cyanobacteria, which reside several metres underwater, cannot receive the correct wavelengths required to cause photoinduced charge separation in conventional photosynthetic pigments. To combat this problem, a series of proteins with different pigments surround the reaction centre. This unit is called a phycobilisome.

Carbon Dioxide Levels and Photorespiration

As carbon dioxide concentrations rise, the rate at which sugars are made by the light-independent reactions increases until limited by other factors. RuBisCO, the enzyme that captures carbon dioxide in the light-independent reactions, has a binding affinity for both carbon dioxide and oxygen. When the concentration of carbon dioxide is high, RuBisCO will fix carbon dioxide. However, if the carbon dioxide concentration is low, RuBisCO will bind oxygen instead of carbon dioxide. This process, called photorespiration, uses energy, but does not produce sugars. RuBisCO oxygenase activity is disadvantageous to plants for several reasons:

1. One product of oxygenase activity is phosphoglycolate (2 carbon) instead of 3-phosphoglycerate (3 carbon). Phosphoglycolate cannot be metabolized by the Calvin-Benson cycle and represents carbon lost from the cycle. A high oxygenase activity, therefore, drains the sugars that are required to recycle ribulose 5-bisphosphate and for the continuation of the Calvin-Benson cycle.
2. Phosphoglycolate is quickly metabolized to glycolate that is toxic to a plant at a high concentration; it inhibits photosynthesis.
3. Salvaging glycolate is an energetically expensive process that uses the glycolate pathway, and only 75% of the carbon is returned to the Calvin-Benson cycle as 3-phosphoglycerate. The reactions also produce ammonia (NH_3), which is able to diffuse out of the plant, leading to a loss of nitrogen.

A highly simplified summary is:

2 glycolate + ATP $\rightarrow$ 3-phosphoglycerate + carbon dioxide + ADP + NH_3

The salvaging pathway for the products of RuBisCO oxygenase activity is more commonly known as photorespiration, since it is characterized by light-dependent oxygen consumption and the release of carbon dioxide.

Crop Nutrition and Fertilizer Requirements

Essential Plant Nutrients

Proper nutrition is essential for satisfactory crop growth and production. The use of soil tests can help to determine the status of plant available nutrients to develop fertilizer recommendations to achieve optimum crop production. The profit potential for farmers depends on producing enough crop per acre to keep production costs below the selling price. Efficient application of the correct types and amounts of fertilizers for the supply of the nutrients is an important part of achieving profitable yields.

There are at least 16 elements known to be essential for plant growth. Carbon (C), hydrogen (H), and oxygen (O) are derived from carbon dioxide (CO2) and water (H2O). Nitrogen (N), phosphorus (P), potassium (K), sulphur (S), calcium (Ca), magnesium (Mg), boron (B), chlorine(Cl), copper (Cu), iron (Fe), manganese (Mn), molybdenum (Mo) and zinc (Zn) are normally derived from the soil in the form of inorganic salts. Ninety-four to 99.5 per cent of fresh plant material is made up of carbon, hydrogen and oxygen. The other nutrients make up the remaining 0.5 to 6.0 per cent.

Macronutrients refer to those elements that are used in relatively large amounts, whereas micronutrients refer to those elements that are required in relatively small amounts.

Table: *Essential Plant Nutrients*

Supplied from air and water	***Supplied from soil and fertilizer sources***	
	Macronutrients	***Micronutrients***
Carbon (C)	Nitrogen (N)	Zinc (Z)
Hydrogen (H)	Phosphorous (P)	Copper (Cu)
Oxygen (O)	Potassium (K)	Iron (Fe)
	Sulphur (S)	Maganese (M)
	Calcium (Ca)	Boron (B)
	Magnesium (Mg)	Chlorine (Cl)
		Molybdenum (Mo)
		Cobalt (Co)

All 13 elements must be present in the soil for plant use, in varying degrees of availability, to ensure both the immediate and long term needs of the crop.

Some of the commonly used terms to describe levels of nutrient elements in plants include:

Deficient	When an essential element is at a low concentration that severely limits yield and produces more or less distinct deficiency symptoms. Extreme deficiencies will lead to death of the plant.
Insufficient	When the level of an essential plant nutrient is below that required for optimum yields or when there is an imbalance with another nutrient. Symptoms of this condition are seldom evident.
Sufficient	When the concentration of an essential nutrient is present in adequate amounts for optimum crop growth.
Excessive	When the concentration of an essential plant nutrient is sufficiently high to result in a corresponding shortage of another nutrient.
Toxic	When the concentration of either essential or other elements is sufficiently high to reduce plant growth severely. Severe toxicity will result in death of plants.

Soil Nutrient Content and Soil Testing

Nitrogen and phosphorus are the most commonly deficient nutrients in Canadian prairie soils. Potassium and sulphur deficiencies occur in particular areas and soil types. Calcium and magnesium are contained in lime which is plentiful in most prairie soils and therefore deficiency problems are rare. Research has found micronutrient deficiency problems are not common on the prairies. However, specific soil conditions have been identified where inadequate levels of micronutrients occur. In central Alberta, on the Black, Gray-Black transition soils and organic soils, Copper (Cu) deficiency problems and significant responses to Cu fertilizer have been observed.

Soil sampling and testing can give an excellent inventory of plant available nutrients and other soil chemical factors important for crop production. This inventory is a basis for recommending additional nutrients for crop production on an individual field basis.

Soil nutrient levels vary from year to year, and frequently will vary within fields, even on fields that seem to be uniform. It is therefore necessary to follow certain recommended steps for soil sampling and testing to develop a sound ongoing soil fertility management programme.

An understanding of general nutrient status can be obtained for a field if soil tests are conducted. Nitrogen soil testing is recommended annually as the available nitrogen can change considerably from year to year. Changes are dependent upon environmental conditions such as rainfall and temperature patterns during a growing season, type and yield of crop harvested, date of harvest, fall tillage, amount of fertilizer applied to the previous crop. Potassium and phosphorus levels do not change substantially in a soil over a period of several years. Therefore, sampling for phosphorus and potassium may be conducted every 2 or 3 years, or when changing crop type. Sampling for sulphur should be done annually, unless a previous test indicates that the available levels to 60 cm (24 in) are in excess of 55 kg/ha (50 lb/ac). In the latter case, a test for sulphur every 2 to 3 years is adequate.

While it is recognized that soil testing is not an infallible guide to crop production and that other factors also come into play, soil tests help to reduce the guesswork in fertilizer practices. In the past, Alberta farmers that have soil tested, have generally applied significantly more fertilizer than farmers that did not soil test. Past observations suggest that Alberta farmers that do not soil test in many instances may be applying less nitrogen fertilizer than needed to obtain the most economical yield.

Poor soil sampling technique is a major problem which causes variation in fertilizer recommendations. Soil testing is only as good as the quality of the soil samples. Therefore, good soil sampling procedures must include:

1. Determine where and how to soil sample each field.
2. Use proper equipment and supplies.
3. Sample at the proper time of the year.
4. Obtain samples from the correct soil depths.
5. Handle soil samples properly.

Being familiar with the proper soil sampling procedures is important whether you are doing the sampling yourself or it is being done by a custom operator.

Determine Where and How to Soil Sample Each Field

Soil sampling should be done on an individual field basis. Samples from different fields should not be mixed. Begin by evaluating each field to determine representative areas. Sample hilly fields with knolls, slopes, or depressions, separatedly from mid slope positions to ensure any potential sulphur or other nutrient problem is detected.

Major areas within fields having distinctly different soil properties such as texture should be sampled and fertilized as separate fields because of different nutrient requirements.

Problem areas such as saline spots, poorly drained depressions, and eroded knolls should not be sampled unless they represent a significant portion of the field. If they do, separate samples should be obtained. In addition, other abnormal areas such as old manure piles, burnpiles, haystacks, corrals, fence rows or farmstead sites should be avoided. At least 15 to 20 sampling sites are required for each field to give a good representative sample. Samples taken from only four or five sites in a field are generally not representative and often result in incorrect fertilizer recommendations.

There are four basic methods for taking soil samples:

Benchmark Soil Sampling - This method involves selecting uniquely different areas within a field and sampling each area separately. Unique areas are selected based on soil types, topography and crop growth. Once sites are selected, the producer will take soil samples from each specific area each year to use as a guide of fertilizing all similar areas within the field. This method is rapidly gaining popularity in Alberta, particularly with farmers that are adopting precision farming techniques. This is the current recommended method which has the best chance to maximize you're economic yield.

Grid Soil Sampling - With this method, a field is sampled in an organized grid pattern. Soil sample frequency may range from taking one sample in 0.5 acre units of the field to one sample for each 5.0 acre units of the field. The smaller the soil sampling unit the greater the accuracy of the sample. The advantage of this method is that a field map that can be prepared for each nutrient and be used for variable rate fertilization and precision farming. The cost of taking the soil samples and the soil analysis is very high and therefore is not economical for many producers.

Topographic Soil Sampling - With this method, a producer selects the separate soil sampling sites based on topography. A set of soil samples is taken from each uniquely different topographic area within a field.

Random Soil Sampling - This involves taking soil samples in a random pattern across a field, generally avoiding unusual or problem soil areas within a field. Generally the field should not be more than 80 acres in size and has been cropped uniformly in the past. Normally, 15 to 20 sites must be sampled to obtain a representative soil sample of the field. This is the most common method of sampling presently used in Alberta.

Use Proper Equipment and Supplies

A soil sampling probe is best for taking samples to the 60 cm (24 in) sampling depth. Use clean labelled plastic pails for collecting samples. Metal pails should not be used if micronutrient testing is to be done. Soil sample augers can also be used but it can be difficult to accurately separate soil samples into 0-15, 15-30 and 30-60 cm depths.

Tools may be borrowed from a fertilizer dealer, your local Alberta Agriculture district office, or may be purchased from some soil testing laboratories or fertilizer dealers. Information sheets, soil sample cartons, and shipping boxes are available from soil testing laboratories.

Sample at the Proper Time of the Year

Ideally, samples should be taken just prior to seeding. However, from a practical standpoint, this is difficult because little time is left to plan a fertilizer programme and purchase fertilizer in time for seeding. The best alternative is to obtain soil samples taken in the fall once soil microbial activity has declined. The proper sample time in the fall is after the soil surface temperature drops to less than 7°C. At this temperature soil processes such as mineralization (breakdown of soil organic matter into plant available nutrients) that cause changes in soil nutrients proceed quite slowly and therefore changes in plant available nutrient levels are normally not great. Generally, it is safe to soil sample in most areas of Alberta after the beginning of October. By sampling in the fall, there is sufficient time to properly process samples, provide test results and recommendations and develop a fertilizer programme for that fall or the next spring.

Obtain Soil Samples from the Correct Depths

Many soil testing labs suggest that a 0 to 30 cm (0 to 12 in) depth sample is adequate for developing fertilizer recommendations. However, for ideal plant nutrient evaluation it is suggested that samples be taken from the 0 to 15 cm (0 to 6 in) and 15 to 30 cm (6 to 12 in), separately and also take samples from the 30 to 60 cm (12 to 24 in) depths. Keep samples from each depth in a separate container.

Phosphorus and potassium recommendations are based on a measure of the amounts of the available forms of each of these nutrients contained in the 0 to 15 cm (0 to 6 in) depth sample. Generally, most of the plant available P in soil is confined to the plow layer as P is very immobile. Nitrate-nitrogen (NO3-N) and sulphate-sulphur (SO4-S) are both mobile nutrients may be found in significant amounts in the 30 to 60 cm depth. Therefore, N and S fertilizer recommendations based on a 0 to 15 cm (0 to 6 in) depth sample usually suggest a higher than necessary fertilizer application rate. For this reason, recommendations regarding nitrogen and sulphur are based on a measure of the amount of nitrogen and sulphur from depth (0-60 cm) soil samples.

Handle Soil Samples Properly

As mentioned, soil from each depth should be placed in separate containers. Immediately after the samples have been taken:

1. Mix the soil in each container thoroughly in order to obtain a homogeneous mixture.
2. Remove the soil and spread on a piece of clean paper.
3. Allow the soil to completely air dry at a temperature of not more than 30°C. Do not dry in an oven or at a high temperature since this can change the levels of some nutrients.

4. Care should be taken to avoid contamination of the samples with fertilizer materials such as commercial fertilizers, manure, salt, water and dust. Samples should not be dried on old fertilizer or feedbags or areas where fertilizers have been handled.
5. A fan may be used to ensure constant air flow over samples to enhance drying.

Once the samples are thoroughly dry, fill the soil sample cartons. Label each carton with correct field number and sample depth. Complete an information sheet on cropping and fertilizer history and mail to a reputable soil testing laboratory. The services of soil testing and fertilizer recommendations are not free. Consult your Alberta Agriculture crop specialist for details of private soil testing laboratories that are available.

Fertilizer Recommendations

Fertilizer recommendations are based on the results of the soil test analyses and on the nutrient requirement of the crop to be grown. Recommendations on time and method of fertilizer application are also included. Each soil testing lab has its own philosophy for making fertilizer recommendations. Two examples are:

1. Recommendations which indicate the nutrient requirements and yield potentials for optimum economic production based one or more moisture conditions of the field.
2. "Target Yield Recommendations" which indicate the nutrient requirements for a range of various lower and higher yield potentials under the same moisture conditions. With this information the producers have the flexibility of selecting a fertilizer application rate or target yield that best suits their individual situation.

Producers must keep in mind that optimum yields of high quality spring wheat can not be obtained without adequate fertilization if the crop to be grown on soils deficient in essential elements. Fertilization, however, will neither increase yield or quality of wheat if other management inputs and cultural practices are not optimal, nor will it increase yield if the added nutrients are not required. Therefore, the most successful fertilizer programme will be based on a knowledge of soil nutrient status combined with optimum crop and fertilizer management practices.

Plant Tissue Analysis

Plant tissue analysis measures nutrient levels in the plant during their growth. The supply of available nutrients is reflected in the nutrient content of the crop. Therefore use of plant tissue analysis allows a producer to evaluate the effectiveness of fertilizer recommendations from a soil testing service. Producers who do not soil test can still use routine plant tissue analysis to evaluate their fertilizer management programme to determine whether they used the correct kinds and amounts of nutrient.

Plant tissue analysis can be used to diagnose crop nutrition production problems. For example, at times, a plant growth problem occurs and can not be explained. Soil, climatic and other environmental conditions seem favourable, essential plant nutrients were supplied, and other sound management practices were followed, plant tissue analysis can indicate if nutrient levels may be associated with the problem.

For reliable results and useful interpretation, producers must follow proper procedures for plant sampling and sample handling. The sample must represent the crop it was taken from. Management decisions based on plant tissue analysis results reflect the quality of the sample provided.

In sampling wheat plants, parts to sample are the entire above-ground plant. Ideally, tissue samples must be taken when the plant is at Zadok's growth stage 30. Growth stage 30 is when the leaf sheath of wheat has begun to lengthen and are strongly erected, but before the first node of the stem is visible at the base of the plant. If the sample is taken prior to growth stage 30, nitrogen is more concentrated in the leaf, so the prescription would call for too little nitrogen to be applied. If taken after growth stage 30, rapid growth has begun and nitrogen content will be diluted by increases in dry matter resulting in excessive nitrogen fertilizer estimates.

An adequate sample must contain 20 to 50 individual plants. Samples should be taken in the morning (or on cool or cloudy days) because heat and moisture stress often occurs during the mid-day and mid afternoons, on hot sunny days or immediately following a rain. Plant samples should be cut with a clean, sharp, rust free knife, blade or scissors. Clean plastic containers should be used for sample collection. In subsequent handling, clean brown paper bag can be used.

After sampling:

1. Remove loose soil particles with a clean, dry cloth or brush.
2. Do not wash the sample.
3. Lay out the sample on clean paper until completely air dried at normal room temperature (25 to 35°C).
4. Do not oven dry samples.

Completely dried samples can then be placed in a plant tissue sample bag or clean brown paper bag. Information Sheets from a testing lab should be completed for each sample as this information helps when interpreting the results. The sample bag and the corresponding information sheet should each be carefully labelled with the same identity so that samples and sheets can be matched in the laboratory.

The laboratory will determine the levels of each nutrient in the plant sample, and will indicate if each nutrient level is excessive, adequate, marginal or below a critical level. The critical concentration of a nutrient is the point where crop growth may be 10 per cent or less than the maximum. This system uses a previously established set of standards for nutrients in a specific plant part, sampled at a particular growing stage and then compared those normals to the findings in the sample. The principle of this system lies in the relationship between nutrient concentration and crop yield.

The most meaningful definition of critical nutrient concentration for efficient growers is the level of a nutrient below which crop yield, quality, or performance is unsatisfactory. In spring wheat, 2.0-3.0% N, 0.26-0.5% P, and 1.5-3.0% K in the whole plant prior to filling are considered sufficient.

Table: *Plant Tissue Analysis Interpretative Critera*

Crop Cereals Plant Part/Growth Stage Whole Plant Prior to Filling

Nutrient		***Low***	***Marginal***	***Sufficient***	***High***	***Excess***
	Spring	1.5	1.5 - 2.0	2.0 - 3.0	3.0 - 4.0	4.0
Nitrogen (N) %	Winter	1.25	1.25 - 1.75	1.75 - 3.0	3.0 - 4.0	4.0
Phosphorous (P) %		0.15	0.15 - 0.25	0.26 - 0.5	0.5 - 0.8	0.8
Potassium (K) %		1.0	1.0 - 1.5	1.5 - 3.0	3.0 - 5.0	5.0
Sulphur (S) %		0.1	0.1 - 0.15	0.15 - 0.40	0.40 - 0.8	0.8
Calcium (Ca) %	Other	0.10	0.10 - 0.2	0.2 - 1.0	1.0 - 1.5	1.5
	Barley	0.20	0.2 - 0.3	0.3 - 2.0	2.0 - 2.5	2.5
Magnesium (Mg) %		0.1	0.1 - 0.15	0.15 - 0.50	0.5 - 1.0	1.0
Zinc (Zn) ppm		10	10 - 15	15 - 70	70 - 150	150
Copper (Cu) ppm	Barley	2.3	2.3 - 3.7	3.7 - 25	25 - 50	50
	Wheat	3.0	3.0 - 4.5	4.5 - 25	25 - 50	50
	Oats	1.7	1.7 - 2.5	2.5 - 25	25 - 50	50
Iron (Fe) ppm		15	15 - 20	20 - 250	250 - 500	500
Manganese (Mn) ppm		10	10 - 15	15 - 100	100 - 250	250
Boron (B) ppm		3	3 - 5	5 - 25	25 - 75	75
Molybdenum (Mo) ppm		0.01	.01 - .02	.03 - 5	5 - 10	10

Source: *Manitoba Provincial Soil Testing Laboratory, 1987.*

The CNC (critical nutrient concentration) system is the commonly used method in interpreting plant tissue analysis in the prairie provinces as well as other parts of the world. However there are some limitations associated with this system. Some of the limitations include:

Low Sensitivity: The diagnosis requires that the nutrient concentration of plant tissue be compared with previously set values, usually in the form of permissible ranges of variation for each element at a particular stage of growth. These ranges are often very wide, and are therefore not very sensitive.

Inconvenience: It is not always convenient to take samples at a particular physiological stage of the plant's life.

Plant Growth Stage: Inadequacy in dealing with the variations in nutrient concentration with the stage of the plant growth and which plant parts that are sampled.

Multiple Growth Stress: Inability to establish the relative importance of the limiting factors in terms of yield when more than one nutrient being diagnosed is deficient. If other stresses such as drought, salinity, diseases or insects are causing plant growth problems, these can affect plant nutrient concentration in the tissue.

However, the critical nutrient concentration system, when used together with soil testing and careful visual observations can be a useful tool for diagnosing crop problems.

Micronutrient Requirements of Crops

Crops require 16 essential elements to grow properly. The elements include carbon (C), hydrogen (H) and oxygen (02), which are derived from air and water. Ail the remaining nutrients used by plants come from soil in the form of inorganic salts. Legumes are an exception because they can also fix nitrogen from the air.

The *macronutrients* obtained from the soil include nitrogen (N), phosphorus (P), potassium (K), calcium (Ca), magnesium (Mg), and sulphur (S). The remaining essential elements needed by plants are known as micronutrients because plants use them in relatively small amounts. They include: boron (B), chlorine (Cl), copper (Cu), iron (Fe) manganese (Mn), molybdenum (Mo) a . nd zinc (Zn). Carbon, hydrogen and oxygen comprise from 94.0 to 99.5 per cent of fresh plant tissue. The remaining nutrients, which come from the soil, make up the balance of the tissue.

The term micronutrient refers to the relative quantity of a nutrient that is required for plant growth. It does not mean that they are less important to plants than other nutrients. Table lists amounts of micronutrients removed from the soil by several crops. Plant growth and development may be retarded if any of these elements is lacking in the soil or is not adequately balanced with other nutrients.

This fact sheet describes where potential micronutrient deficiencies may occur in Alberta, how to determine if a micronutrient is deficient and how to correct the deficiency.

Sources of Micronutrients in Soils

Inorganic micronutrients occur naturally in soil minerals. The parent material from which the soil developed and soil forming processes determine what the micronutrient content of the soil will be. As minerals break down during soil formation, micronutrients are gradually released in a form that is available to plants. Two sources of readily available micronutrients exist in soil: nutrients that are adsorbed onto soil colloids (very small soil particles) and nutrients that are in the form of salts dissolved in the soil solution.

Organic matter is an important secondary source of some micronutrients. Most micronutrients are held tightly in complex organic compounds and may not be readily available to plants. However, they can be an important source of micronutrients when they are slowly released into a plant available form as organic matter decomposes.

Table: *Amounts of some micronutrients removed by good yields of various crops.*

Crops harvested and portion used for analysis		*Micronutients removed (kg/ha)*						
		Yield level t/ha	*Chlorine (Cl)*	*Boron (B)*	*Copper (Cu)*	*Iron (Fe)*	*Manganese (Mn)*	*Zinc (Zn)*
Alfalfa	- hay	1.3	6	0.10	< 0.1	0.20	0.70	0.70
Barley	- grain	4.0	8	0.10	< 0.1	0.30	0.10	0.10
	- straw	-	1	0.02	< 0.1	0.01	0.70	0.10
Corn	- grain	9.5	2	0.70	< 0.1	0.20	0.10	0.20
	- stover	-	1	0.06	< 0.1	1.0	1.70	0.30
Oats	- grain	4.0	1	-	< 0.1	1.00	0.20	0.10
	- straw	-	1	-	< 0.1	0.20	0.20	0.40
Peas	- vines & pods	-	-	0.07	< 0.1	0.70	0.50	0.10
Potatoes	- white, tubers	40	27	0.07	< 0.1	0.90	0.20	0.10
Wheat	- grain	4.0	6	0.06	< 0.1	0.50	0.20	0.20
	- straw	-	2	0.02	< 0.1	0.20	0.30	0.10

Data compiled from several sources.

Soil Factors That Affect Micronutrient Availability

Physical and chemical characteristics of soil affect the availability and uptake of micronutrients:

- Soils low in organic matter (less than 2.0%) may have lower micronutrient availability.
- Soils with higher amounts of clay (fine texture) are less likely to be low in plant available micronutrients. Sandy soils (course texture) are more likely to be low in micronutrients.
- Soils that have very high levels of organic matter (greater than 30% organic matter to a depth of 30 cm) often have low micronutrient availability.
- Soil temperature and moisture are important factors. Cool, wet soils reduce the rate and amount of micronutrients that may be taken up by crops.
- As soil pH increases the availability of micronutrients decreases, with the exception of molybdenum.

Determining the Need for Micronutrients

Diagnosing a micronutrient deficiency can be a difficult and time consuming process. To identify a micronutrient deficiency follow these steps:

1. Ensure that poor crop growth is not the result of a macronutrient deficiency, drought, salinity, disease or insect problem, herbicide injury or some physiological problem.
2. Find out if a micronutrient deficiency has been identified before in a particular crop or soil type in the area.
3. Examine the affected crop for specific micronutrient deficiency symptoms.
4. Take separate soil samples from both the affected and unaffected areas for complete analysis, including micronutrients.
5. Send plant tissue samples from both the affected and unaffected areas for complete analysis that includes tests for micronutrient levels.
6. If all indications point to a micronutrient deficiency, apply the micronutrient to a specific, clearly marked out affected area of land to observe results in subsequent seasons.

Potential Micronutrient Deficient Areas in Alberta

While Alberta soils generally contain adequate amounts of micronutrients, some micronutrient deficiencies have been documented.

As soils continue to be cropped, micronutrient deficiencies may become more common as available levels of some elements are depleted.

Boron

Boron deficiencies have been suspected in canola and alfalfa grown on sandy-textured Grey Wooded soils. However, research specifically documenting the response to added boron is limited.

Brown and Dark Brown irrigated soils in southern Alberta will frequently test deficient for boron. However, cereal crops do not respond to additions of boron. Canola, pea and bean yields have declined by 1 0 to 20 per cent due to boron toxicity after a 2 lb/ac application of boron.

Chlorine

Chlorine is not known to be deficient in Alberta soils from a nutritional standpoint. The requirement of plants for chlorine is satisfied by the chlorine found in the soil and received from rainfall. Chloride is also added to soil in potash fertilizer (KCl). Research has shown that chloride added at rates higher than required to meet nutritional needs is associated with suppression of root and leaf diseases in some cereal crops. The reasons for disease suppression in some soils are poorly understood.

Copper

Research has shown that cereal crops grown on organic soils (greater than 30% organic matter to a depth of 30 cm) often respond to copper fertilization. More recently, copper deficiency has been identified in wheat, barley and oats grown on mineral soils in the Black and Grey-Black soil zones of Alberta. Copper deficient soils tend to be either sandy or light loam soils with relatively high levels of organic matter (6-10%). High levels of soil phosphorus or heavy applications of manure are often associated with a copper deficiency on these soils.

Wheat, barley and oats are the most sensitive to a copper deficiency. Park spring wheat and Condor barley are the varieties that are the most sensitive to copper deficiency and show the most obvious disease symptoms. Rye and canola are relatively tolerant to a copper deficiency.

Iron

Iron deficiencies have not been observed in field crops in Alberta. However, iron deficiency symptoms such as leaf yellowing are common among various trees, shrubs and ornamentals on high pH soils because lime reduces the availability of iron.

Manganese

Manganese deficiencies are most common on organic soils and high pH mineral soils. Deficiency symptoms are commonly observed following cool wet conditions in spring. Oats are more susceptible to a manganese deficiency than other cereal crops. Organic soils with a high pH are the most likely to respond to manganese fertilizer.

Molybdenum

Molybdenum deficiencies have not been diagnosed in field crops in Alberta. However, isolated deficiencies have been observed in vegetable crops such as cauliflower.

Zinc

Zinc deficiencies tend to occur on calcareous, high pH soils that have been machine levelled, are sandy in texture or have relatively high soil phosphorus levels' Deficiencies are most common in spring when conditions are cool and wet. In southern Alberta, irrigated field beans have responded to applications of zinc. Zinc deficiencies have been suspected in some irrigated corn fields in southern Alberta, but research trials have not confirmed this.

A response to added zinc may occur on badly eroded soils or soils that have had large amounts of added phosphate fertilizer.

Micronutrient Deficiency Symptoms

Some micronutrients have characteristic deficiency symptoms. However, symptoms can be easily confused with other nutrient deficiencies, salinity, disease, drought, herbicide injury or other physiological problems. Visual symptoms alone are not a reliable method of determining a micronutrient, problem, but they are useful indicators when used with other diagnostic tools.

Boron

This deficiency results in stunted growth of young plants. The youngest leaves are affected first. They will be misshapen, thick, brittle and small. Because boron is not easily transferred from old to young leaves, older leaves usually remain green and appear healthy. Often dark brown, irregular lesions appear, followed by pale yellow chlorosis of young leaves. Stems are short and growing points may die. In canola, the symptoms of a boron deficiency can be confused with a sulphur deficiency. In alfalfa, boron deficiency symptoms include death of the terminal bud, rosetting, yellowtop and poor flowering.

When a boron deficiency is moderate, seed yield is often reduced without any evidence of severe deficiency symptoms during vegetative growth.

Chlorine

Chlorine deficiencies are very rare; therefore, symptoms are seldom observed. Symptoms may include stubby roots, some chlorosis of new leaves and plant wilting.

Copper

Copper is not readily transferred from old to young leaves, so older leaves remain darker and relatively healthy and the deficiency symptoms develop on younger leaves. The visual symptoms of a copper deficiency in wheat include yellowing of younger leaves, limpness, wilting, pigtailing (whip tailing or curling) of the upper leaves and kinking of the leaf tips. Other signs include excessive tillering, aborted heads, delayed maturity and poor grain filling resulting in a high straw to grain ratio. On copper deficient soils these symptoms tend to occur in irregular patches. Copper deficiency is often associated with the disease stem or head melanosis and an increased incidence of ergot. For barley, the symptoms of a copper deficiency include yellowing, pigtailing, awn kinking, excessive tillering and weak straw. Oats will also show pigtailing. Copper deficiency symptoms have not been well documented for canola or alfalfa.

Iron

Chlorosis of the younger leaves characterizes an iron deficiency. The tissue between the veins gradually turns yellow, while the veins tend to stay green. The tips and margins of some leaves may turn brown and become dry and brittle.

Manganese

In legumes, deficiency symptoms include pale green young leaves and a pale yellow mottling develops in interveinal areas, while the veins remain green. Oats are an excellent indicator crop. Manganese is partly mobile in oats. White to grey flecks or specks first appear and become more severe on mature leaves about halfway up the shoot. If a deficiency persists, symptoms spread to old leaves then to the youngest leaves. The specked condition is referred to as "grey speck" and will appear in the interveinal area of the lower half of older leaves and extend toward the tip as symptoms develop.

Manganese is not readily transferred from old to young leaves in wheat and barely. In wheat and barley, affected young leaves frequently turn pale green and have a limp or wilted appearance. A mild interveinal chlorosis develops in the mid-section of the leaf and spreads rapidly becoming pale yellow-green. Small white to grey spots, specks or strips appear a short distance from the end of the leaf tip on young leaves.

Molybdenum

Molybdenum deficiency symptoms are similar to those of nitrogen. Since molybdenum deficiencies are very uncommon symptoms are rarely seen.

Zinc

Zinc is partly mobile in wheat and barley. In these crops pale yellow chlorotic areas appear on middle leaves, halfway up the stem. Chlorotic symptoms first develop in the lower half or mid-section of the leaf followed by grey or dark brown necrosis of the leaf. Generally, stems are very short and often fan-shaped with leaves crowded together at the top.

Zinc deficient beans are stunted and older leaves are smaller and narrower. The older leaves may have light blotches between the veins. Younger leaves will have a more normal healthy green colour but may be smaller.

In flax, a zinc deficiency can cause grayish-brown spots in the younger leaves with shortened internode spaces and stunted appearance.

In corn, symptoms occur within a few weeks of emergence as light yellow bands on the youngest leaves. The most severe symptoms occur on the youngest leaves from the unfolding bud, referred to as "white-bud". Old leaves remain dark green and appear healthy. In a prolonged case of deficiency the middle leaves develop pale yellow interveinal chlorosis near the tips. A zinc deficiency prevents the elongation of internodes and leaves, which results in short stems with the leaves crowded together at the top in a fan-shaped appearance.

Note that zinc deficiency symptoms are similar to those of manganese and iron in some crops.

Soil Sampling and Testing

Soil tests aid in determining whether a particular nutrient is responsible for poor production and provides the basis for deciding the type and amount of fertilizer needed to correct a nutrient shortage. A soil sample used for laboratory analysis must consist of a composite of a number of samples taken from the field. The DTPA (diethylenetriamine pentaacetic acid) method is used to extract the metal micronutrients. Hot water is used to extract boron. Either water or sodium nitrate is used to extract chlorine. Analytical results are usually given in parts per million (ppm). No soil test has proven particularly useful in predicting the availability of molybdenum. The micronutrients that are normally tested for are boron, copper, iron, manganese, and zinc.

Surface soil samples (0-1 5 cm) should be taken from both the affected area and an adjacent area of good crop growth for comparison. The general range levels used for determining when to add micronutrients for improving crop production. When a soil sample tests low in a micronutrient, a potential micronutrient deficiency may occur. Some soils with low micronutrient levels at the surface (0- 1 5 cm) do not respond to fertilization because they have higher levels of the nutrient in the subsoil.

Micronutrient Application in Test Strips

Even after observing micronutrient deficiency symptoms and conducting soil and plant tissue analyses it still can be difficult to predict if a profitable yield response will occur. Therefore, the micronutrient suspected of being deficient should be applied to a small affected area of the field in a carefully marked test strip. Visual observations and yields from the treated and untreated areas should be taken to determine if a measurable response occurred.

Boron

Application of borate of borax can be either broadcast and incorporated or banded. Boron containing fertilizers should not come into contact with the seed at planting time. For permanent crops such as alfalfa, boron should be used as a topdressing. Soil application rates should not exceed 1.5lb/ac on soils with a pH less than 6.5 to avoid boron toxicity problems. Foliar applications should not exceed 0.3lb/ac to avoid toxicity problems. For all types of applications, extreme care must e taken to apply the correct amount and to apply it uniformly to avoid toxicity problems.

Copper

Broadcast and incorporated rates of 3 to 7 lb/ac of copper in the form of copper sulphate or copper oxide is recommended for deficient mineral soils. On organic soils, broadcast and incorporated rates of 10 to 15 lb/ac are required. Soil application rates should be effective for up to 10 years. Chelated forms of copper are also very effective in the year of application but little is known about the residual effects in Alberta soils. Foliar application on mineral and organic soils is not as consistent but can be used after deficiency symptoms appear. Foliar applications are required annually and are most effective at the late tillering stage. If the deficiency is severe, two applications (mid-tillering and boot stage) are necessary. Foliar rates of between 0.1 to 0.3 lb/ac are recommended.

Manganese

Only limited information is available on manganese fertilization in Alberta. As a rule, broadcast applications are seldom effective. For cereals, a seed placed treatment of manganese sulphate should be most effective. Foliar application can also be used if deficiency symptoms develop during the growing season.

Zinc

For treatment of a zinc deficient soil where a sensitive crop such as beans or corn is grown, a band application of 2 to 5 lb/ac of zinc sulphate or 0.5 to 1.0 lb/ac of a chelated zinc is recommended. When zinc deficiencies are suspected early in the growing season, a foliar application of 0.5 lb/ac of zinc sulphate can be used. Severely, deficient beans may require two applications. On eroded soils a 5]b/ac broadcast and incorporated application of zinc sulphate can be tried.

Conclusions

The need for micronutrients in crop production have long been recognized in Alberta. Zinc deficiencies in irrigated beans in southern Alberta were detected in the early 1980s. Research by Alberta Agriculture and Agriculture Canada has clearly shown the need for copper fertilizer particularly for wheat grown on organic soils and on some Black and Grey-Black soils in central Alberta. One million acres may be copper deficient in central Alberta. Some suspected manganese deficiencies are presently being investigated. Continued research and testing is required to identify when various crops will give economic yield increases to additions of micronutrient fertilizers.

It is important to keep the need for micronutrient fertilizers in perspective. Over-promotion of micronutrients has occurred on occasion. Some farmers have applied micronutrients in the hope of increasing crop yields even though there is little evidence to suggest a deficiency exists.

Farmers who are concerned about micronutrient deficiencies are encouraged to investigate the need thoroughly and apply the nutrients in test strips if necessary. The test strip areas must be carefully marked out for comparison to areas where micronutrients were not used. Visual and qualitative comparisons should be made on these test strips.

There is no question that micronutrient levels will gradually decline in soils over time as cropping continues. However, most Alberta soils are currently well supplied with micronutrients. Soils and crops in Alberta that require micronutrient fertilizers are the exception not the rule. Care must be taken to keep the need for micronutrient fertilizers in perspective and not to promote them beyond their true significance.

Bibliography

Andrews J.: *The Domesticated Capsicums*, University of Texas Press. 1995.

Arthur W. Gilbert, Mortier F. Barrus and Daniel Dean: *Growing and Breeding of Potatoes*, Asiatic Pub, Delhi, 2006.

Banki, L.: *Bioassay of Pesticides in the Laboratory,* Akademiai Kiado, Budapest, 1978.

Barbeau G.: *Tropical Fruits in Nicaragua*, Managua, Nicaragua Ministerio de Desarrollo Agropecuario, Agraria, 1990.

Berger, P.L.: *Pyramids of Sacrifice: Political Ethics and Social Change*, New York, Basic Books, 1974.

Burr, Fearing Jr.: *The Field and Garden Vegetables of America*, Timber Press, Portland, 1988.

Chadha K. L. and Pareek O. P.: *Advances in Horticulture: Fruit Crops,* New Delhi, Malhotra Publishing House, 1993.

Cronquist, A.: *The Evolution and Classification of Flowering Plants*, New York Botanical Garden, Bronx, New York., 1988

Currah L. and Proctor F. J.: *Onions in Tropical Regions*, Kent, Natural Resources Institute, 1990.

Daniells J.: *Illustrated Guide to the Identification of Banana varieties in the South Pacific*, Canberra, ACIAR, 1995.

Dharamvir Hota: *Modern Biotechnology in Plant Breeding*, Gene Tech Books, Delhi, 2007.

Diederichsen A.: *Coriander (Coriandrum sativum L.).* Rome, International Plant Genetic Resources Institute, 1996.

Fellows, P.J.: *Food Processing Technology*, Woodhead Publishing, Cambridge, UK, 1993.

Ferentinos L.: *Proceeding of the Sustainable Taro Culture for the Pacific Conference*, Honolulu, HITAHR, 1993.

Fortuner, R.: *Nematode Identification and Expert System Technology,* New York, Plenum Press, 1988.

Georges, S.: *The Debt Boomerang: How Third World Debt Harms Us All*, Boulder, Westview Press, 1992.

Godden G.: *Growing Citrus Trees*, Australia, Lothian Publishing Company Pvt. Ltd., 1988.

Hardy B.: *Biology and Agronomy of Forage Arachis*, Cali, International Centre for Tropical Agriculture, 1994.

Herminie Broedel Kitchen: *Soils and Crops : Diagnostic Techniques*, Satish Serial Publishing, Allahabad, 2004.

Jean, F. C.: *Unpublished Data, Department of Botany,* University of Nebraska, 1925.

Keshav Prasad Yadav: *Application of Morphometry in Geomorphology*, Radha Pub, Delhi, 2008.

Lorenzen, S.: *The Phylogenetic Systematics of Freeliving Nematodes*, London, The Ray Society, 1994.

Madulid Domingo A.: *A Pictorial Cyclopedia of Philippine Ornamental Plants*, Philippines, Makati Metro Manila, 1995.

Malins A.: *Postharvest Handling of Pineapple and Mango*, Port of Spain, Trinidad and Tobago, 1992.

Mitra S.: *Postharvest Physiology and Storage of Tropical and Subtropical Fruits*, Oxon, CABI, 1997.

Nobel, P. S.: *Physicochemical and Environmental Plant Physiology*, Academic Press, San Diego, 1999.

Oka H. I.: *Origin of Cultivated Rice*, Elsevier, Japan Scientific Societies Press, 1988.

Pemberton, R. W.: *Predictable Risk to Native Plants in Weed Biological Control,* Oecologia, 2000.

Pramila Tripathi: *Botanical Pesticides in the Management of Post-harvest Fruit Diseases*, Daya, Delhi, 2005.

Qystein V. Sjaastad: *Physiology of Domestic Animals*, International Book Distributing Co., Delhi, 2005.

Roper, Elinor M. C.: *Seedtime, The History of Essex Seeds*, Phillimore & Co. Ltd., Chichester, 1989.

Rutherford Lyn.: *A Gourmet's Book of Mushrooms & Truffles*, Sydney, Golden Press Pvt. Ltd., 1991.

Sanchez, P. A.: *Properties and Management of Soils in the Tropics*, John Wiley and Sons, New York, 1976.

Thomas E.: *Fruit Production in St. Kitts and Nevis*, Port of Spain, IICA, 1996.

Yoshida T.: *Cultivation of Citrus Genetic Resources for Evaluation of Characteristics*, Tokyo, JICA, 1996.

Index

❑❑❑